Adolf Grünbaum

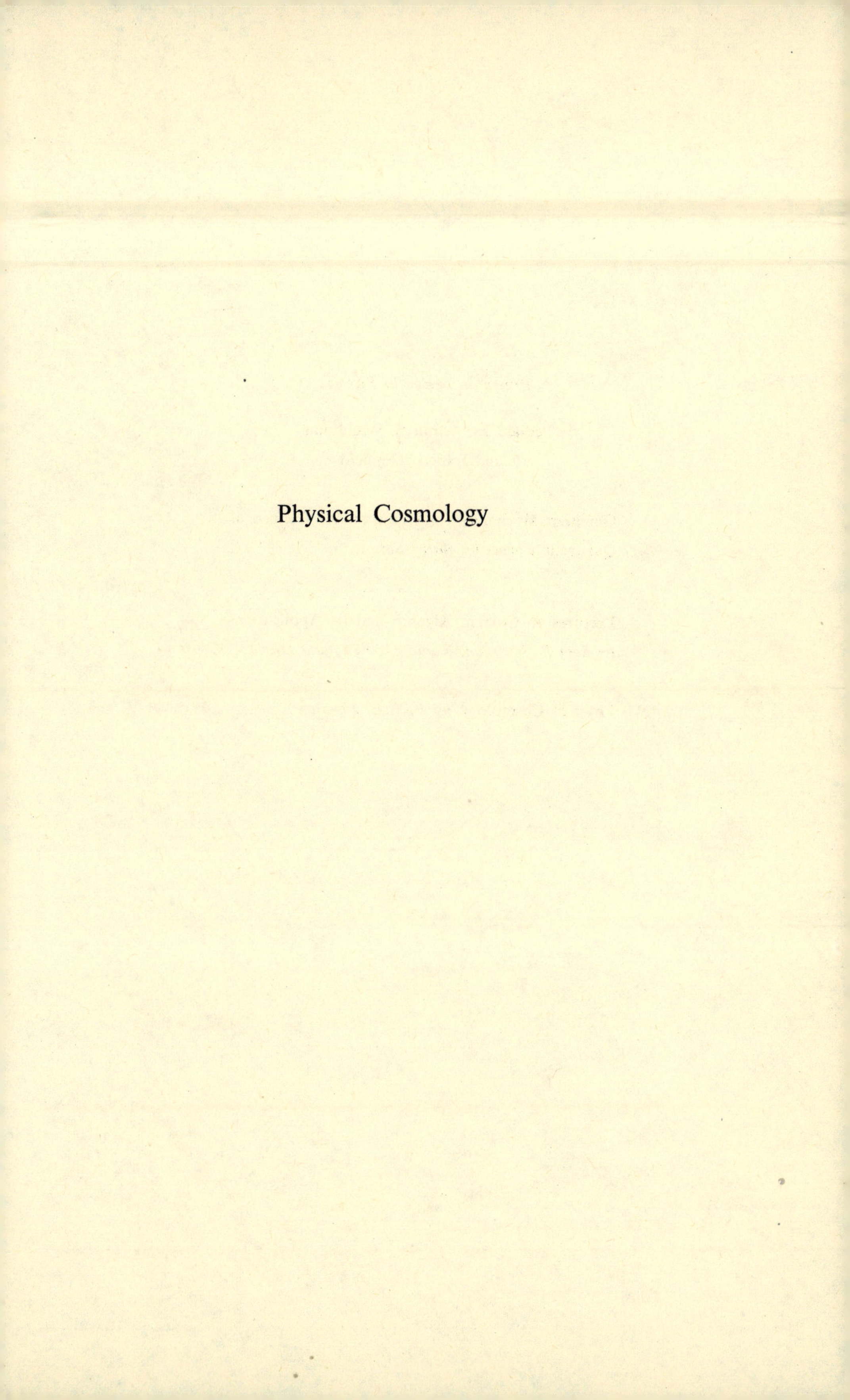

Physical Cosmology

Princeton Series in Physics

edited by Arthur S. Wightman
and John J. Hopfield

Quantum Mechanics for Hamiltonians Defined as Quadratic Forms *by Barry Simon*

Lectures on Current Algebra and Its Applications *by Sam B. Treiman, Roman Jackiw, and David J. Gross*

Physical Cosmology *by P. J. E. Peebles*

Physical Cosmology

BY P. J. E. PEEBLES

Princeton Series in Physics

Princeton University Press
Princeton, New Jersey · 1971

L.C. Card: 74-181520

ISBN 0-691-08108-5 (paperback edn.)

ISBN 0-691-08137-9 (hardcover edn.)

Second Printing, 1974

Printed in the United States of America
by Princeton University Press

To Alison

INTRODUCTION

Physical science more often than not violates the rule that familiarity breeds contempt. We may be assured that we have a reasonable understanding of the atmosphere of Mars, the major constituents, the temperature and pressure, yet we know very well that it is hard to say whether it will rain tomorrow. This does not mean that we understand the Martian atmosphere better than our own. Quite the contrary, we have incomparably better observational material on the Earth's atmosphere so we can and should ask much more detailed and deeper questions about it. Sometimes this order may be bent because greater physical simplicity can overweigh better data. For all the wealth of evidence on the center of the Earth from seismic data, free oscillations, heat flux, and gravitational anomalies, to say nothing of hard rock geology, it could be that we understand the center of the Sun better than the center of the Earth just because the ideal gas law is a good approximation for the Sun.

In cosmology the reliance on physical simplicity, pure thought and revealed knowledge is carried well beyond the fringe because we have so little else to go on. By this desperate course we have arrived at a few simple pictures of what the Universe may be like. The great goal now is to become more familiar with the Universe, to learn whether any of these pictures may be a reasonable approximation, and if so how the approximation may be improved. The great excitement in cosmology is that the prospects for doing this seem to be excellent. On one front we have some explicit questions, like the shapes of the redshift-magnitude relation and the redshift-angular size relation, and the values of some elementary parameters like Hubble's constant and the mean mass density. Probably we will only know for sure whether these questions are clever after we understand the answers, but at least we can see how available technological developments can be turned to powerful new attacks on them. A second exciting front is the search for new phenomena that may or may not interconnect in a pleasant way with accepted ideas. Of course we have no plan for finding these nuggets, only the expectation based on past form that as techniques of observation develop and extend our view we are liable to stumble on them.

The goal of these notes is to aid this advance of cosmology by providing a guide, mainly for the experimentalist or observer, to the popular questions and answers in the subject. While there is no shortage of questions it is true that I have restricted the list to the orthodox questions in an establishment cosmology. There is the point of view that in a science as primitive as cosmology one ought to avoid orthodoxy, give equal time to all competing cosmologies. The first problem is that one could never hope to think of all reasonably possible views people might want to consider. Second there is the danger that the approach will degenerate into long lists of alternative possibilities, from which the reader can understand where we are only with the help of an auxiliary guide. My own preference is to make a subjective selection of a reasonably possible cosmology, and then study it in all the detail one can muster. The end result of such an effort may be to reveal that the cosmology is logically inconsistent or even in conflict with observation, which is progress of a sort. The hope is that the development of the observations may convince us of the rough validity of the chosen cosmology, and guide us to how the cosmology should evolve.

Consistent with this view the Cosmological Constant Λ is seldom mentioned in these notes because a non-zero Λ can do only two essentially different things. If Λ is less than zero, or positive but not too large, it alters somewhat the connection among expansion rate, mass density and curvature in the present Universe, and it hardly at all affects the expansion rate in the early Universe. We gain one more parameter, but we already had more than we can fix. The second possibility is more interesting – if Λ were large enough it would eliminate the Big Bang, the singular infinite density epoch in the models. This is a serious change in the picture, and the consequences deserve to be followed up in detail to see how well it all hangs together. However, it will be not be done here. Another perturbation to the establishment is the scalar-tensor gravity theory of Carl Brans and Robert Dicke. This affects the expansion rate only very slightly now, but could very markedly increase the expected expansion rate in the early Universe. Again, the first effect is a detail among details so far as the over-all nature of the Universe goes, the second an interesting branch of ideas that should be pursued, but not here.

The ordering of material in the notes requires some explanation. The first chapter is an introduction to the elements of classical cosmology, written in the standard pseudo-historical style. That is, the main points are set forth roughly in the order in which they were discovered or came to the general attention of the scientific world, along with the names to which the discoveries conventionally are credited. The survey covers the span of time from Slipher's first measurement of the redshift of an extragalactic nebula, in 1912, through the formulation of the Steady State model by Bondi, Gold and Hoyle, in 1948. This pseudo-historical approach is useful because it efficiently communicates the accepted views. The great disadvantage is that it obscures the other branches in the road, including those that merit closer attention. It is hoped, however, that even with the rough places planed the tentative nature of the subject will be apparent.

Each of the following chapters is a detailed discussion of a separate topic in cosmology. Chapter II deals with the observational basis for the assumption that the Universe is homogeneous and isotropic in the large scale average. Chapter III is devoted to the scale of the Universe as provided by the empirical value of Hubble's constant H, and to the comparison of this measure with other indications of the time scale for the Universe. Chapter IV is a lengthy discussion of the mean mass density ρ. This is a fascinating topic because we are required to make a catalogue of all the significant contents of the Universe, a subject that seems likely to keep us all gainfully employed for a long time. It will be noted that the relative length of chapters, as on ρ and H, is a measure of what I could think to say about them, not a judgment about relative importance or richness of the topic.

Chapter V is a discussion of the microwave background as a candidate for the Primeval Fireball. Chapter VI is devoted to the theoretical basis of the general relativity cosmological models, and to the classical tests of these models. It may seem very strange that the theoretical treatment of the models is only given here, not in the first or second chapter. The reason is that almost all the preceding discussion of basic parameters can be understood in a simple way, using at most symmetry arguments, and it seems wise not to burden the fragile observational evidence with an unneeded theoretical superstructure. The theory is needed for an explicit discussion of the classical tests like the redshift-magnitude relation, but clearly even here we should be worrying about the empirical test, not about the superstructure per se.

Chapter VII deals with possible histories of the Universe. This task looks to be of a greater magnitude of difficulty even than the questions discussed in the earlier chapters, and, since we are not doing all that well on those "easier" problems, it has been argued that it would be a waste of time to try anything more ambitious. This may be true, but of course we will only find out if we take the plunge – formulate a picture and exhaustively discuss its observational consequences. This is an enormous task, and the best that can be done is to report on some preliminary dips.

Chapter VIII is devoted to helium production in the Big Bang. This is a peculiar subject. Under a reasonably short list of explicit assumptions we arrive at the sharp prediction that the earliest stars should have formed out of primeval material containing helium in the amount of about 25 percent by mass. Depending on how the observations go this may prove to be a miraculously powerful probe of the early Universe, or it may prove to be irrelevant because a very specific chain of events did not happen.

Finally, there is a short Appendix explaining two standard astronomical measures, parsec and magnitude, for the benefit of physicists. There is also a list explaining the more common symbols used in the notes.

These notes are based on a graduate course on cosmology I gave in the Physics Department in Princeton in the fall of 1969. The original notes contained a discussion of how galaxies may have formed, but this has been almost entirely eliminated from the present version, the exception being some effects described in Chapter VII that could be treated in a non-relativistic Newtonian approximation. I hope to discuss in a separate volume the very broad problem of galaxy formation, and the presumably related task of understanding irregularities of all sorts in an expanding Universe.

In writing these notes I have been strongly guided, directly and indirectly, by many colleagues, most particularly by the members of the experimental gravity research group that condensed around R. H. Dicke at Princeton. These notes were commenced under the direct prodding of J. A. Wheeler. I should like to explicitly thank colleagues who read and helped with parts of the notes: Jon Arons, R. H. Dicke, D. Hawley, D. Morton and David Wilkinson. I am grateful to Alison Peebles and Marion Fugill for their assistance.

CONTENTS

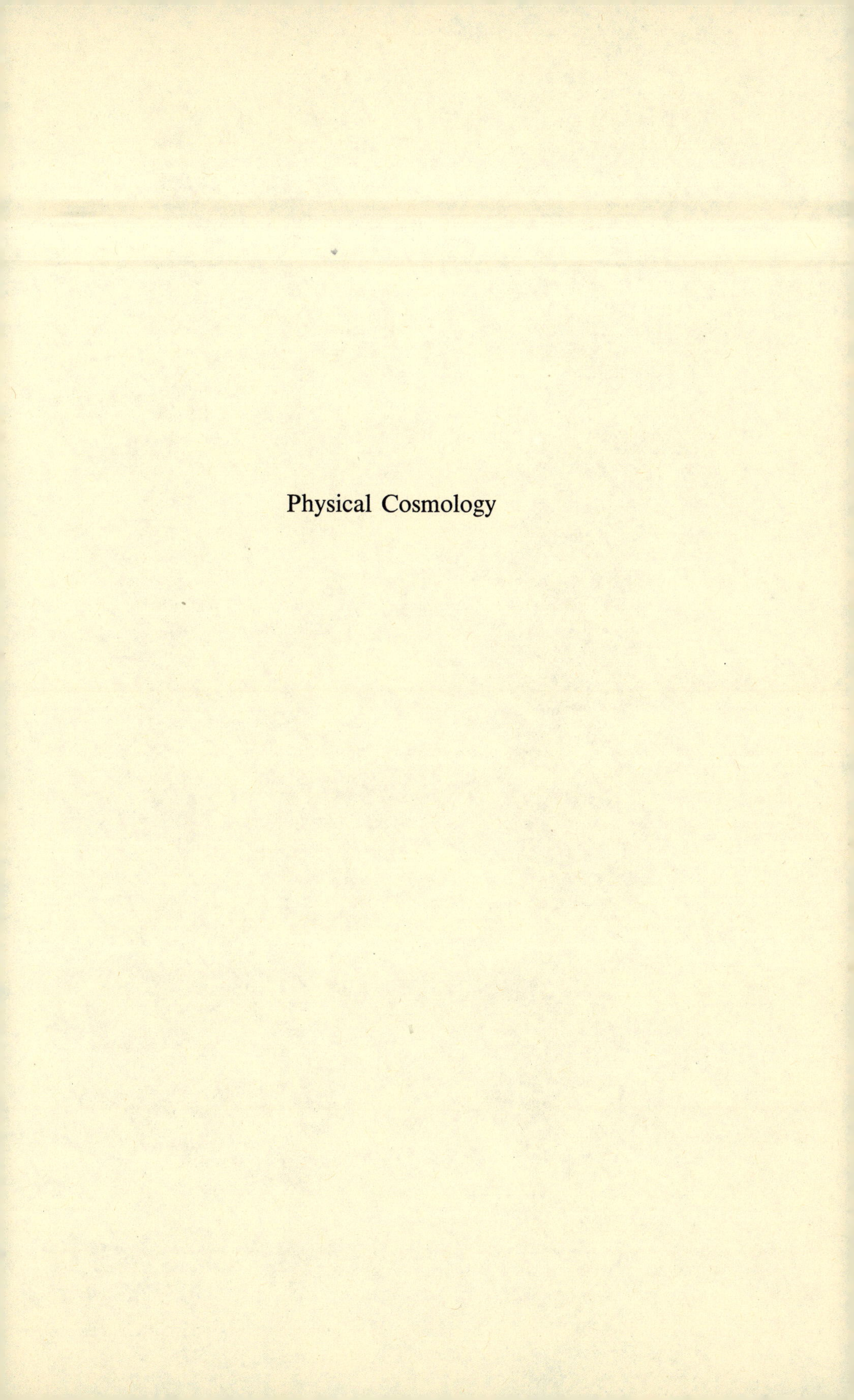

Physical Cosmology

I. GOLDEN MOMENTS IN COSMOLOGY 1912-1950

a) *The Expansion of the Universe*

The first observational basis of modern cosmology was the discovery of the nature of the extragalactic nebulae, or galaxies, and of their distribution and motion. Hubble gives an admirable account of this discovery in *Realm of the Nebulae.*[1]

The apparent radial motion of the nebulae is indicated by the observed frequency shift in their spectra. In 1912 Slipher obtained the first positive measurement of this effect for an extragalactic nebula, the Andromeda Nebula, M 31. He found a velocity of approach of 300 km sec^{-1}, consistent with the best modern value.[2] Slipher continued painstaking accumulation of data on apparent radial velocities of the nebulae, and by 1924 Eddington could list in his *Mathematical Theory of Relativity* apparent velocities obtained by Slipher for 41 objects.[3] Of these 36 are positive, that is, the observed spectrum is shifted toward the red, as if the nebulae were moving away from us.

These data found their way into a book on relativity theory because of a suspicion that the tendency toward apparent recession of the nebulae may have something to do with the cosmological model invented by deSitter in 1917.[4] DeSitter followed the two assumptions Einstein had adopted in his cosmological model:[5] a) The Universe is homogeneous and isotropic, so that the Universe appears the same in any direction or from any spot; b) The Universe is unchanging, the mean mass density is constant and the curvature of space is constant. Einstein found that his original general relativity theory would not admit a solution consistent with these two conditions. Therefore he added to his gravitational field equations a new long-range effect, the Cosmological Constant Λ term, which with suitable adjustment of the parameters would serve to balance the gravitational attraction among matter and make a static model. DeSitter observed that he could obtain another solution with constant non-zero curvature simply by removing all the matter from the model. The solution is stationary just because there is no matter to move about. If now some test particles were introduced in this DeSitter model the Λ term would cause the particles to "scatter," or accelerate away from each other. It is this effect that people thought might have some connection with the observed recession of the nebulae. The theoretical Sturm und Drang on the subject of the Λ term has been chronicled by North.[6]

All of this discussion was highly uncertain because people could not be sure whether the spiral nebulae were separate "island universes," star systems like our own Milky Way Galaxy, or perhaps only lesser objects swarming around the Galaxy. It was known from star counts as functions of apparent magnitude that the Galaxy is a finite flattened star system, thought to be about centered on the Solar System – this picture was called the Kapetyn universe.[7] This was well accepted by 1917, so Einstein's assumption that the system of stars about us extends out without boundary was clearly and literally false, but it is today thought to be accurately valid when applied not to stars but to the large-scale mean distribution of galaxies. This brilliant coup by Einstein was not entirely a good thing, however, for it helped inspire later cosmologists similarly to attempt to deduce the nature of the Universe from pure thought, with at best questionable results.

The first step in the discovery of the nature of the spiral nebulae was Shapley's determination (in about 1918) of the distance scale in the Galaxy.[8] Shapley made use of Cepheid and RR Lyrae variable stars, for which the luminosity is a periodic function of time. It was known from the variable stars at very nearly common distance from us in the Magellanic clouds that the period of variability is correlated with absolute magnitude (total luminosity, ergs sec^{-1}). This enables one to determine the relative distances of different Cepheid variables from their periods and apparent magnitudes (brightness in the sky, ergs cm^{-2} sec^{-1}). To fix the absolute distances (determine the zero point of the period-luminosity relation) Shapley used the method of statistical parallaxes, which had been applied earlier by Hertzsprung and others. There are some nearby Cepheid variables for which one has both proper motions (rate of angular motion across the sky) and radial velocities. If the radial and transverse velocities are the results of random motions of the stars then evidently the average radial motion (km $\sec^{-1}$) and proper motion (radians $\sec^{-1}$) ought to be in proportion to the absolute distance to the star (km). Shapley determined the distances to globular clusters (dense nearly spherical star clusters) by isolating in these systems Cepheid-like variables. He argued that the globular clusters form a sort of halo around the Galaxy, and, because most of the globulars are concentrated in one part of the sky, that we are well away from the center of the system. Shapley concluded that the center of the Galaxy is some 20 kpc away. This number has now been reduced to around 10 kpc, the main correction being for interstellar absorption.

The distances to the nearer spiral nebulae were convincingly established with the discovery of Cepheid variable stars in these systems. This was accomplished with the help of the 100" telescope on Mount Wilson, completed in 1918. Apparently the first Cepheid variable identified in an extragalactic nebula was in M 31, the Andromeda Nebula, the nearest giant galaxy and the other dominant member of the Local Group of galaxies. In the Annual Report of the Mount Wilson Observatory for 1924 it was announced that Hubble had found several variable stars in M 31, at least the brightest of which had a light curve typical of the Cepheid variables. Interestingly enough the report did not state the distance to M 31 this would imply ($\sim$ 300 kpc according to Shapley's calibration of the period-luminosity relation) perhaps because in the same Report there is tentative confirmation of van Maanen's suggestion that the spiral nebulae may exhibit appreciable proper motions of rotation or streaming along the spiral arms.[9] The suggested proper motion was $\sim 3 \times 10^{-3}$ arc sec/year. This with the observed internal radial velocities within the nebulae indicated that they were typically 1 kpc away, well within the Galaxy. Subsequent developments proved that this was in error.

The next year Hubble presented data on variable stars in another object, the irregular nebula NGC 6822.[10] Eleven stars had light curves typical of Cepheids. The observed periods and apparent magnitudes were related in the standard way, so Hubble had every reason to suppose that these stars were ordinary Cepheid variables, but at unusually great distance. From Shapley's normalization of the period-luminosity relation based on galactic Cepheids Hubble concluded that the distance to NGC 6822 is 214 kpc (this later was increased by a factor of 2, as described in Chapter III). Hubble pointed out that this put the nebula well outside the Galaxy, and made it about comparable in linear extent and luminosity to the Magellanic Clouds, which are satellites of the Galaxy. It is interesting that two years earlier Shapley had arrived at the same conclusion from morphological arguments.[11] He argued that NGC 6822 looks rather similar to the Magellanic Clouds but the angular size is smaller, and that star clusters in NGC 6822 look like those in the Magellanic Clouds but smaller and dimmer, and finally that one can apparently distinguish some bright individual stars in NGC 6822 but they are fainter than the brightest stars in the Magellanic Clouds. Each of these points suggested to Shapley that NGC 6822 is about 300 kpc away. There seemed to be good reason, therefore, to believe that NGC 6822 is an independent star system, roughly on a par with the Milky Way Galaxy, although smaller in radius by a factor of perhaps 40.

Having the distances to some nearby galaxies from the Cepheid variables, Hubble could find the distances to other galaxies with the help of another indicator, the brightest star in the galaxy. The brightest stars seem to have approximately constant absolute magnitude, and they are a more powerful distance indicator just because they can be picked out at greater distance. Hubble calibrated the absolute magnitude of the brightest star in the few galaxies to which he had distances from Cepheid variables. For the larger sample of galaxies in which he could pick out what seemed to be individual bright stars he formed an estimate of the typical absolute magnitude of a galaxy, which he could then use to carry his distance estimates to still more remote galaxies.

In 1929 Hubble published his discovery that the apparent recession velocity of a galaxy, as determined from Slipher's observations of the shift of the spectral lines, and corrected for the peculiar velocity of the Solar System (mainly, this is the rotation velocity about the center of the Galaxy), is directly proportional to his estimate of the distance of the galaxy,[12]

$$v = H\ell, \qquad H \cong 500 \text{ km sec}^{-1} \text{ Mpc}^{-1}. \tag{1}$$

The constant of proportionality H (Hubble used K) is now called Hubble's constant, and as described in Chapter III the best modern value is a factor 5-10 lower than (1).

The linear connection between redshift and distance was anticipated by some theoretical work prior to 1929. In 1923 Weyl[13] had remarked that in a deSitter model the motion of a set of test particles with appropriately chosen initial conditions will at any later time satisfy a linear relation of the form (1) when ℓ is small enough. Robertson rediscovered this in 1928,[14] and he even remarked that Slipher's redshift data taken together with the data Hubble was accumulating on distances to the nebulae roughly verify the linear expression. In *Measure of the Universe* North suggests that Hubble was influenced by Robertson in arriving at his law of the general recession.[6] Certainly Robertson was at Caltech in 1927-1929, and Hubble was aware of the "deSitter effect" because he mentions it in an uncharacteristically theoretical last paragraph of his paper. However, in view of Hubble's healthy distrust of theorists (as expressed for example in *Realm of the Nebulae*), I would be surprised if he had in mind anything more than the simplest reasonable expression of his results.

The deSitter model with appropriate initial values for the test particles is a special limiting case of the evolving homogeneous and isotropic cosmological models. These models all imply the relation (1), where in general H could be positive or negative (galaxies receding or approaching). The evolving models were derived independently by Friedman[15] and Lemaître.[16] In Lemaître's discussion the line element in comoving coordinates (that is, each galaxy has fixed spatial coordinates r, θ, ϕ) may be written as

$$ds^2 = dt^2 - \frac{a(t)^2 dr^2}{1-r^2/R^2} - a(t)^2 r^2 (d\theta^2 + \sin^2\theta d\phi^2) \tag{2}$$

where R is a constant and the function $a(t)$ now is called the expansion parameter. On substituting the indicated components of the metric tensor into Einstein's gravitational field equations, Lemaître obtained a differential equation for the expansion parameter,

$$\frac{\dot{a}^2}{a^2} = \frac{8}{3}\pi G\rho(t) - \frac{c^2}{R^2 a^2} + \frac{\Lambda}{3}, \tag{3}$$

where the mass density ρ includes the mass equivalent of any energy present (like electromagnetic radiation), and Λ is Einstein's Cosmological Constant. Lemaître's second differential equation needed to specify the cosmological model is

$$\frac{d\rho}{dt} + 3(\rho + \frac{P}{c^2})\frac{\dot{a}}{a} = 0. \tag{4}$$

He correctly identified this as the expression of local energy conservation, the term $3P\,\dot{a}/a$ representing the work done per unit volume against the pressure P of the fluid.

Lemaître was clearly aware of the observational significance of his evolving cosmological model. He recognized that the galaxies might map out the large-scale structure of the Universe. He derived the linear relation (1) when ℓ is small, and he stated without reference a value for the constant of proportionality, $H = 600$ km sec^{-1} Mpc^{-1}. This is so close to Hubble's value that there must have been communication of some sort between the two. Although the theory of the evolving universe clearly was anticipated by Friedman, it was Lemaître who had the great good fortune to have derived the expanding cosmological model at the right time, when the basic phenomenon, the law of the general recession of the galaxies, was just becoming clarified, and Lemaître recognized the significance of this phenomenon. According to the usual criterion for establishing credit for scientific discoveries Lemaître deserves to be called the "Father of the Big Bang Cosmology."

The dramatic agreement of theory and observation in the expression (1) was considered strong evidence in favor of Lemaître's picture. Thus Einstein promptly advised that his original cosmological model be abandoned, and with it the Cosmological Constant he had introduced for the purpose of constructing the model.[17] There was very soon some discussion of possible interconnections of the supposed expansion of the Universe with other phenomena, as is the proper course in a physical science. Before describing this let us consider a few subsequent points of clarification of the theory of the Lemaître cosmological model.

b) *Nature of the Lemaître Model*

Robertson showed that, in a general relativity world picture assumed to be homogeneous and isotropic, the most general line element is the expression (2) with the constant R^{-2} allowed to be positive or negative or zero.[18] In this expression the coordinates are comoving, that is the spatial coordinates are tied to the matter. Later (1935) Robertson and Walker independently demonstrated that in *any* homogeneous and isotropic world model based on a Riemannian geometry the expression (2), with the constant R^{-2} and the function $a(t)$ freely chosen, is the most general possible line element (cf. Sec. VI-a-i). For this reason (2) is known as the Robertson-Walker line element.

There is a direct connection between Hubble's law and the assumed homogeneity and isotropy of the Universe. According to Robertson and Walker homogeneity and isotropy imply the line element (2). The line element fixes the proper time or distance interval (as measured with real rods and clocks) between two points (events) in space-time in terms of the coordinate interval between the points. For example, if two points are separated by the interval $dt = d\theta = d\phi = 0$, $dr > 0$, the proper distance between the points is $d\ell = a\,dr\,(1 - r^2/R^2)^{-1/2}$. We can fix the origin of coordinates on any chosen galaxy, like our own. Then the proper distance to another galaxy at coordinate distance $r << |R|$ is

$$\ell(t) = a(t)\,r. \tag{5}$$

Because the radial coordinate r of a chosen galaxy is fixed, the coordinates being comoving, the velocity of recession of the galaxy relative to our own is

$$v = \frac{d\ell}{dt} = \dot{a}r = \frac{\dot{a}}{a}\ell = H\ell, \qquad H = \frac{\dot{a}}{a}, \tag{6}$$

which is Hubble's law.

Apparently Milne first pointed out that the form of Hubble's law is an immediate consequence of the assumed homogeneity and isotropy of the Universe.[19] He showed this by an even simpler argument, which in modified form may be stated as follows. Let us consider three galaxies separated by distances so large that local irregularities like the Local Group may be ignored, but so small that we can ignore relativistic effects in the expansion. If the Universe is expanding in a homogeneous and isotropic way the triangle defined by the three galaxies must at all times remain similar to the original triangle. This means that the length of each side has to scale by the same factor, say $a(t)$, as the Universe expands. By extending the net to a fourth galaxy, a fifth, and so on, we see that $a(t)$ has to be a universal scale factor (when local irregularities are ignored). Thus the distance between two galaxies satisfies

$$\ell(t) = \ell_0 \, a(t), \tag{7}$$

where ℓ_0 is independent of time. Hubble's law follows as before (eq. 6).

Milne and McCrea followed up this line of argument to obtain in very simple fashion the differential equation for $a(t)$ from ordinary Newtonian mechanics.[20] Again in somewhat modified form the argument runs as follows.[21] Imagine with Lemaître that the galaxies are considered as a gas of particles, and that the Universe is so large that this gas is adequately represented as a continuous fluid with mass density ρ. It will be assumed for the moment that this gas has negligible pressure. Now imagine that at some place there is drawn a spherical volume, radius ℓ, and that all the matter within the sphere is temporarily removed and set to one side. What will be the curvature of space within the evacuated sphere? The answer is a generalization of Newton's theorem that within a hollow iron sphere the gravitational field due to the sphere vanishes. The analogous statement in general relativity is that within a hollow centrally symmetric system space is *flat*. This is a trivial application of Birkhoff's theorem.[22] When the matter is put back in the sphere it is therefore added to flat space. One can always choose the radius ℓ so small that the dimensionless number $G\mathfrak{M}/\ell c^2 << 1$, where $\mathfrak{M}$ is the mass within ℓ; but when this condition is satisfied (and $\Lambda = 0$) matter added to flat space obeys the laws of Newtonian mechanics! This means that we can treat the motion of our sphere in the Newtonian approximation, as if it were an isolated sphere of matter, and the radius of the sphere evidently satisfies the familiar equation

$$\frac{d^2\ell}{dt^2} = -\frac{G\mathfrak{M}}{\ell^2}, \tag{8}$$

where the mass $\mathfrak{M}$ of the sphere is of course constant. On multiplying this expression by $\dot{\ell}$ we get

$$\frac{d}{dt}\frac{\dot{\ell}^2}{2} = -\frac{G\mathfrak{M}\dot{\ell}}{\ell^2} = \frac{d}{dt}\frac{G\mathfrak{M}}{\ell},$$

or

$$\dot{\ell}^2 = \frac{2G\mathfrak{M}}{\ell} + K, \tag{9}$$

where K is the constant of integration. Finally, with the help of (7) we can rewrite (9) as

$$\frac{\dot{a}^2}{a^2} = \frac{8}{3}\pi G\rho - \frac{c^2}{R^2 a^2}, \tag{10}$$

where the constant has been rewritten as R^{-2} to conform with the usual notation. This will be recognized as Lemaître's equation (3) when $\Lambda = 0$.

This is a perfectly correct derivation, not a Newtonian analog.[21] It is true that we used the condition $G\mathfrak{M}/\ell c^2 << 1$, but this can be made arbitrarily good by making the constant ℓ_0 in equation (7) small enough, and it will be noted that ℓ_0 cancels out of (10).

When pressure cannot be neglected, as in a model containing substantial amounts of radiation, one must invoke another result of General Relativity, that in the weak field limit the active gravitational mass of a fluid is

$\rho + 3P/c^2$.[23] Here ρc^2 is by definition the net energy of the fluid (including annihilation energy) per unit volume, and P is the fluid pressure, both evaluated in the fluid rest frame. Thus to take account of the active gravitational mass due to pressure equation (8) should be rewritten as

$$d^2\ell/dt^2 = -\frac{4}{3}\pi G(\rho + 3P/c^2)\ell,$$

which with (7) gives

$$\ddot{a}/a = -\frac{4}{3}\pi G(\rho + 3P/c^2). \tag{11}$$

As Lemaître remarked, equation (4) follows from energy conservation: the total energy $\mathfrak{M}c^2$ within the sphere of radius $\ell(t)$ decreases as ℓ increases because the fluid within the sphere is doing pressure work against the exterior,

$$\begin{aligned} \frac{d}{dt}\mathfrak{M}c^2 &= -P\frac{dV}{dt} = -4\pi\ell^2 P\frac{d\ell}{dt}, \\ \mathfrak{M} &= \frac{4}{3}\pi\rho\ell^3. \end{aligned} \tag{12}$$

These equations with (7) immediately reduce to (4).

McCrea showed that equation (11) can be integrated once by the following trick.[23] Use equation (4) to eliminate P from (11). The result is a first order linear differential equation in $\rho(t)$, and the standard solution to a differential equation of this sort yields equation (10), where again c^2R^{-2} is the constant of integration.

To understand the behavior of matter and radiation in these expanding models it is useful to consider a comoving observer, an observer at rest relative to the matter about him. This comoving observer sees an isotropic redshift law of the form (1), while an observer moving relative to him looking in one direction would see that the nebulae are receding at a larger rate, and looking in the opposite direction, that the nebulae are receding at a smaller rate, or even approaching. It might be noted that the possibility of defining a preferred motion for the comoving observers does not violate any principle of relativity, for we have only defined the motion relative to neighboring matter.

Consider a free particle moving past a comoving observer at speed v. The particle will be said to have peculiar velocity, or proper velocity, v. If the particle is freely moving, no collisions and no applied electromagnetic or other fields, then the peculiar velocity decreases as the Universe expands, which may be seen as follows. At time t the particle passes the position of a comoving observer who determines that the peculiar velocity is $v(t)$. A time interval dt later the particle has gone a distance vdt, to the position of a second comoving observer. According to Hubble's law the second observer is moving away from the first at the rate (eq. 6)

$$dv = [v(t)\, dt]\, \dot{a}/a,$$

so the peculiar velocity seen by the new observer is

$$v(t + dt) = v(t) - dv = v(t)\,(1 - \dot{a}/a\, dt).$$

The solution to this equation is

$$v(t) \propto a(t)^{-1}. \tag{13}$$

This result should not be interpreted as some sort of cosmic drag or tired matter effect. It is entirely and simply a kinematic effect of the expansion – the peculiar velocity of a freely moving particle decreases because it is always overtaking observers moving away from it.

If the Universe contains a uniformly distributed gas of particles, and if the particles are not created or destroyed, then the particle number density seen by a comoving observer satisfies

$$n(t) \propto a(t)^{-3}. \tag{14}$$

This is because $a(t)$ is the scale factor by which the distance between comoving observers varies and the same scale factor clearly applies to the distance between particles.

Consider next a pulse of radiation ("photon") which passes a comoving observer at time t and is observed to have frequency $\nu(t)$. The discussion leading to (13) may be applied here also. The photon velocity of course remains equal to c but, by the first order Doppler effect, the frequency seen by comoving observers decreases with time. In the time interval dt the photon moves a distance cdt, to a second observer moving away from the first at the rate $(cdt)\,\dot{a}/a$, so the first order Doppler reduction in observed frequency is

$$\nu(t+dt) - \nu(t) = -\nu v/c$$
$$= -\nu d\dot{a}/a\,.$$

The solution to this equation is

$$\nu(t) \propto a(t)^{-1}. \tag{15}$$

This is the general law for the time variation of radiation frequency. The frequency shift of the radiation from a distant galaxy usually is specified by the redshift z defined by the equation

$$1+z = \frac{\nu_e}{\nu_0} = \frac{\lambda_0}{\lambda_e}\,, \tag{16}$$

where ν_0, λ_0 represent the observed frequency and wavelength of a line in the observed spectrum of the galaxy, and ν_e, λ_e are the frequency and wavelength of the radiation as it was emitted by the distant galaxy, as determined by measuring the corresponding spectral line in the laboratory. By (15) and (16)

$$(1+z) = a_0/a(t), \tag{17}$$

where radiation is emitted at epoch t and observed at present epoch t_0, and $a_0 \equiv a(t_0)$. When the redshift is small z is the fractional increase in wavelength, and by the first order Doppler effect zc is the effective apparent recession velocity.

c) *Interconnections – Galaxy Counts, the Time Scale, and the Mean Mass Density*

As soon as the possible connection between the general recession of the nebulae and the expanding Universe model of Lemaître was recognized there was discussion of possible interconnections with other phenomena. As in any physical science the more diverse the interconnections the more credible the theory. Two applications immediately come to mind, in short because one can form from Hubble's constant H a characteristic time H^{-1} and a characteristic density H^2G^{-1}. There is in addition a beautiful test of the homogeneity assumption.

If the galaxies are on the average uniformly distributed through space then the number of galaxies brighter than f (ergs/cm^2 sec) varies as $f^{-3/2}$.

To see this suppose for the moment that all galaxies have the same luminosity $\mathcal{L}$ (ergs sec^{-1}). Then by the inverse square law the galaxies at brightness f are at distance ℓ such that

$$f = \mathcal{L}/4\pi \ell^2 .$$

If the number density of galaxies is n the number of galaxies brighter than f is

$$N(>f) = \frac{4\pi}{3}\ell^3 n = \frac{4\pi}{3} n \left(\frac{\mathcal{L}}{4\pi f}\right)^{3/2} \propto f^{-3/2} . \quad (18)$$

Galaxies have a range of intrinsic luminosities, but (18) holds separately for each class of galaxies with a given absolute luminosity, so by summing over classes evidently the same functional dependence on f holds for the counts of galaxies of all sorts. The very pleasant feature of (18) is that it directly relates observable quantities. It had been known for some time that star counts do not obey (18), and this had been used to estimate the boundary of our star system. Shapley and Ames found rough agreement with (18) in counts of galaxies in their catalogue of the brighter galaxies in the sky.[24] In a massive effort Hubble extended the test down to counts of very faint galaxies in small patches of the sky.[25] This was the first serious attempt to test for the relativistic corrections at high redshift predicted in the Lemaître model, which would for example cause a systematic deviation from (18) when the counts reach faint enough galaxies (cf. Sec. VI-c). Unhappily the test now is considered unreliable because of systematic errors in the measurement of the brightness of dim galaxies – the dimmer galaxies are redder because they are further away, hence redshifted, so there were very serious corrections for the blue-sensitive plates of the day. However, it remains one of our best tests of the large-scale homogeneity of the Universe out to a redshift that is estimated to be $z \sim 0.4$ (Chapter II).

In 1932 Einstein and deSitter proposed that, in the absence of any very secure evidence to the contrary, it may be most reasonable to adopt the simplest assumption that in equation (3) the constants Λ, R^{-2} and the pressure P all are negligibly small.[26] This with equation (6) then implies that the mean mass density is

$$\rho_c = \frac{3H^2}{8\pi G} \sim 5 \times 10^{-28} \text{ g cm}^{-3}, \qquad (19)$$

according to equation (1). This density has a simple Newtonian interpretation. Consider a homogeneous sphere of matter, radius ℓ, expanding in a homogeneous and isotropic fashion (that is, the expansion velocity at distance x from the center is Hx). One can readily verify that if the density of the sphere is ρ_c the material just has escape velocity, the kinetic energy of expansion just amounting to the magnitude of the gravitational potential energy. Evidently if the density were larger than ρ_c the sphere eventually would stop expanding and collapse. This carries over to the Lemaître models with $\Lambda = 0$. If $\rho < \rho_c$ equation (10) says that $R^{-2} < 0$, so $\dot{a}^2$ always is greater than zero and the Universe will expand forever. If $\rho > \rho_c$, equation (10) says $R^{-2} > 0$, and as the first term on the right hand side of (10) decreases at least as fast as a^{-3} (eqs. 4, 14), $\dot{a}^2$ must fall to zero when the expansion parameter gets large enough. At this point the expansion stops and the Universe collapses again. If $R^{-2} < 0$ the model is said to be open, if $R^{-2} > 0$ the model is closed, while if $R^{-2} = 0$ the model is said to be cosmologically flat (cf. Sec. VI-b).

Now it is interesting to ask how (19) compares with the mean mass density in the Universe. By 1932 there were rough estimates of the mean mass density due to the galaxies, the standard game being to find the galaxy number density from the galaxy counts and an estimate of the "typical" absolute luminosity of a galaxy, and then multiplying this number by the "typical" mass of a galaxy estimated from the observed rotation velocities in the outer parts of spiral galaxies. The results were[24,25,27]

$$\rho(G) \sim 10^{-30} \text{ g cm}^{-3}. \qquad (20)$$

The discrepancy between (19) and (20) was not considered serious because, in addition to the substantial uncertainty in (20), one could not be sure how much dark matter there may be between the galaxies. Einstein and deSitter also remarked that if R^{-2} is not negligibly small one can determine its value from equations (6) and (10) once H and ρ are sufficiently well known.[25] As is discussed in Chapter IV we are still far from able to estimate the mass density well enough to do this.

The rough numerical coincidence of (19) and (20) is not strong evidence in favor of any one particular cosmological model. For example, in the model Lemaître worked out in his original paper one assumes that the Universe expands smoothly away from Einstein's static solution. Here the greater the expansion the greater the influence of the Λ term, so one expects to find the numerical coincidence of $G\rho$ and H^2 only if we are near the point of departure from the Einstein case, and in fact Lemaître used rough estimates of ρ and H to determine how far we might have expanded away from the starting quasi-equilibrium radius. Despite these questions, however, it is a magic thing that the characteristic mass density formed from Hubble's constant and Newton's constant agrees in rough order of magnitude with the observed mean mass density due to galaxies, and it does suggest that we may be right in interpreting equation (1) as a significant global property of the Universe rather than as a "tired light" effect, for example.

The second characteristic quantity is H^{-1}, which is roughly the time for the distance between galaxies to double assuming the Universe is in fact expanding. A more definite time depends on the cosmological model. In the Einstein-deSitter model by definition $\Lambda = R^{-2} = P = 0$. Because the pressure is negligible equation (4) says $\rho \propto a^{-3}$, and (10) may be written as

$$a\,\dot{a}^2 = \frac{8}{3}\pi\, G\rho\, a^3,$$

where the right hand side is a constant. Evidently then $a \propto t^{2/3}$, where the time t is measured from zero at the singular point $a = 0$, $\rho \to \infty$. By (6),

$$t = 2/(3\,H)\,. \tag{21}$$

A second interesting model is the low density case, $\rho << \rho_c$ (eq. 19), and $\Lambda = 0$. Here one finds

$$t \cong H^{-1}\,. \tag{22}$$

This is the maximum value for the time since the singularity at $a = 0$, if $\Lambda = 0$.

In the event $\Lambda \neq 0$ the interesting equations are (3) and the equation obtained by differentiating (3) and using (4),

$$\frac{\ddot{a}}{a} = -\frac{4}{3}\pi G\left(\rho + \frac{3P}{c^2}\right) + \frac{\Lambda}{3}. \qquad (23)$$

If $\Lambda < 0$ it makes the curve $a(t)$ steeper in the past than it otherwise would have been, hence reduces the time since the singularity. If $\Lambda > 0$ it can be arranged that the Universe enjoys a quasi-static phase when $\dot{a} \sim 0$, $\ddot{a} \sim 0$. If $P = 0$ equations (3) and (23) give at this phase

$$\Lambda = c^2/R^2 a_s^2 ,$$

which is Einstein's model. Lemaître's original solution is asymptotic to this quasi-static situation in the distant past, which Eddington emphasized is an important point because it shows that Einstein's model is unstable – once set expanding through any chance irregularity it will do so at an ever increasing rate because as a increases $\Lambda/3$ increasingly dominates $\frac{4}{3}\pi G\rho$ in (23).[28] In this picture the age of the Universe depends on how symmetric we are willing to suppose it originally was, but H^{-1} does give a reasonable estimate of the time since the Universe broke away from the quasi-static phase, for we know from the rough agreement of ρ_c and $\rho(G)$ that Λ does not entirely dominate $G\rho$. One can also imagine that the Universe expanded from infinite density, passed through a quasi-static epoch with $\dot{a} \sim 0$, $\ddot{a} \sim 0$, then started expanding again. This picture has been discussed in some detail by Lemaître.[29] On another extreme we can imagine that the Universe originally was contracting, that the contraction eventually was stopped and reversed by a positive Λ term in (23).

In all these models H^{-1} is a significant time scale. If $\Lambda \leq 0$ we expect that stars and galaxies should be younger than H^{-1} because the naive model predicts a singular infinite density state at a time $\leq H^{-1}$ in the past. On the other hand, if we are willing to assume $\Lambda > 0$, we need not expect this strict inequality.

According to equation (1) $H^{-1} \sim 2 \times 10^9$ y. It was promptly recognized that this is an interesting number which should be compared with other "cosmic" ages.

By 1930 there were three interesting radioactive decay ages. It was known that in a uranium ore the abundance of lead relative to uranium is correlated with the geological epoch (geologic age on a relative, not absolute, time scale) of the ore. It was concluded that the lead is mainly the accumulated end product of the uranium decay over the life of the ore. Knowing the uranium decay rate one could therefore fix the age of the ore from the lead-uranium abundance ratio. By 1927 the greatest age established in this way was 1.3×10^9 y.[30] Second, one could estimate the age of the Earth's lead by asking how long it would take for the known mean abundance of uranium in the surface rocks of the Earth to produce by radioactive decay the known mean abundance of lead. The answer was 2 to 6×10^9 years, depending on the assumed whole rock abundances of these elements.[31]

The third age is based on the abundance of the uranium isotope 235.[32] In 1929 Aston identified for the first time the lead isotope 207 in a lead sample chemically extracted from a uranium ore. It was by this time known that the uranium isotope U^{238} is the progenitor of the radium series of of activities ending in Pb^{206}, and that Th^{252} decays to Pb^{208}. It appeared that there must be a uranium isotope U^{235} which is the progenitor of the actinium series ending in the new lead isotope Pb^{207}. Rutherford then proceeded to find the half-life of U^{235}. The relative rates of disintegration of uranium isotopes into the actinium series and the radium series was known to be $K \cong 0.03$. If λ_1 and λ_2 are the decay rates of U^{235} U^{238} then the uranium isotopic abundance ratio satisfies

$$K = \frac{\lambda_1}{\lambda_2} \frac{U^{235}}{U^{238}} .$$

Aston estimated the isotopic abundance ratio in the radiogenic lead to be

$$K' = Pb^{207}/Pb^{206} \sim 0.07 .$$

By the usual laws of radioactive decay,

$$\frac{Pb^{207}}{Pb^{206}} = \frac{U^{235}(e^{\lambda_1 t} - 1)}{U^{238}(e^{\lambda_2 t} - 1)},$$

where t is the age of the ore, $t \cong 10^9$ y. These equations with the known decay rate of U^{238} yield the isotopic abundance ratio $U^{235}/U^{238} \cong 0.003$, and a U^{235} half-life of 4×10^8y.

Rutherford pointed out that if uranium were formed with equal isotopic abundances of U^{235} and U^{238} the abundance ratio would decay down to its present value ~ 0.003 in 3×10^9 y. According to the systematics of isotopic abundances the even-even isotope U^{238} should be formed in no less abundance than U^{235}, so this is expected to be an upper limit to the age of the Earth's uranium. Rutherford compared this age 3×10^9 y, not with the Hubble expansion time H^{-1} but with the age of the Sun, then thought to be as much as 10^{13} y. He concluded that the Sun must have produced uranium along with the planets at a relatively recent time.

There were two main arguments for such large star ages. First, Jeans suggested that the motion and distribution of stars show evidence of partial relaxation to energy equipartition of the sort expected in a gravitating ideal gas in statistical equilibrium. The time to effect this relaxation would be on the order of 10^{13} y.[33] Second, in 1924 Eddington showed that one could give a natural account of the relation between stellar mass and luminosity if it were accepted that matter inside stars obeys the ideal gas law.[34] This theory gave a direct relation between stellar mass and luminosity. The problem was that there are two classes or sequences of stars, giants and dwarfs, the former being accounted for only after the discovery that thermonuclear fusion must be the source of stellar energy and the recognition that the fusion process would in time make a star chemically inhomogeneous. At the time, however, it was thought that the difficulty of the two stellar classes might be relieved if it were supposed that stars can evolve by losing a good fraction of their mass, and the natural way to accomplish this is through the mass equivalent of the energy radiated, the energy coming from the presumed annihilation of electrons and protons. The natural time scale for stellar evolution would be the time for the Solar luminosity to radiate away the equivalent of a Solar mass, and this works out to be $\sim 2 \times 10^{13}$ y.

Very soon after learning of Lemaître's work Eddington remarked on the rather close agreement of H^{-1} and radioactive decay ages, and the substantial disagreement with accepted stellar ages.[28,35] He was inclined to adopt the shorter time scale as the valid one, although it appears that he did not find the choice an easy one. On the other hand, deSitter pointed out that the stars in fact could have existed for 10^{13} y, spending most of this time in a contracting phase of the Universe.[36] This contraction would have turned around some 2×10^9 y ago, perhaps because growing irregularities turned the motion around, perhaps because of the cosmic repulsion effect of the Λ term. He also remarked that the rough agreement of H^{-1} and radioactive decay ages might come about because at the point of maximum contraction the higher density would have favored close stellar collisions that might initiate formation of planetary systems. DeSitter's general point is a good one – it may be very wrong to extrapolate the present apparent expansion of the Universe back in time all the way to infinite density.[37] As discussed in Chapter V the best evidence in favor of this naive extrapolation appears to be the Primeval Fireball radiation.

In his 1937 book on cosmology McVittie concluded: "The shortness of the (cosmological) time scale thus remains an outstanding problem of cosmological theory."[38] However, at that very time it was becoming clear that the source of stellar energy must be the fusion of hydrogen to form helium, not annihilation of matter.[39] Thus the net energy available to a star was reduced from the annihilation energy mc^2 to the packing fraction $0.007\ mc^2$, and the maximum lifetime of the Sun correspondingly was reduced by two orders of magnitude. The lifetime was reduced by another factor of 10 with the recognition that the Sun could remain in its present state only until it had burned out a helium core amounting to perhaps 10 percent of the Solar mass.

In 1946 Bok could conclude that the evidence clearly favored a cosmic time scale of 3 to 5×10^9 y.[40] The energy equipartition arguments had lost much of their appeal when considered in the light of the much more detailed knowledge one had of galactic structure. Bok noted that there remained apparent inconsistencies, but that they were (and are) right on the border of our understanding, where systematic errors and misunderstandings might well be expected. Despite such problems one significant point seemed clear – the stellar evolution ages, radioactive decay ages and Hubble expansion age all pointed to some sort of cosmic event some 3 to 5×10^9 y ago.

d) *The Steady State Cosmology*

By 1948 the Lemaître picture had received some confirmation, but only of a general sort, so it is not surprising that alternate cosmological models were proposed. Of these by far the most influential is the Steady State cosmology.[41]

Bondi and Gold placed a good deal of emphasis on Einstein's assumption that the Universe is in the average homogeneous and isotropic, which by this time had been enshrined by Milne as the Cosmological Principle. They proposed to generalize it to the Perfect Cosmological Principle, that on the average the Universe is invariant against translation in the time direction as well as translation and rotation in spatial directions. A number of results of direct observational importance follow from this assumption in a simple and elegant way. For example, the spatial symmetry allows us to conclude as before that the distance between us and another galaxy satisfies

$$\dot{\ell} = H\ell .$$

In this picture Hubble's constant truly is a constant, so the solution to this equation is

$$\ell \propto e^{Ht} . \qquad (24)$$

Because the galaxies are moving away from each other matter must be spontaneously created to preserve constant nucleon density n, as demanded by the Perfect Cosmological Principle. Consider a comoving volume $V(t)$, expanding with the Universe, so that there is no net flow of matter across the surface of V. By eq. (24) V has to vary as

$$V \propto e^{3Ht} .$$

If nucleons were not created, evidently the nucleon number density would decrease as

$$\frac{\dot{n}}{n} = -\frac{\dot{V}}{V} = -3H .$$

Thus it is necessary to postulate continuous creation at the rate

$$C = 3\,Hn \tag{25}$$

per unit volume.

Any chosen galaxy can have any age, larger or smaller than H^{-1}, but we can find an expression for the mean age of the galaxies within a large volume. Consider again the comoving volume $V(t)$, and let $\mathfrak{N}$ be the mean number density of galaxies. This is related to the rate of creation of galaxies by the same argument that led to (25). Galaxies in V with age t were created when V was smaller than it is now by the factor e^{3Ht}. By (25) the number of galaxies with age t in the interval dt in V is

$$dN = 3H\mathfrak{N}\, V_0 e^{-3Ht}\, dt .$$

Therefore the mean age of the sample of galaxies is

$$\begin{aligned} \langle t \rangle &= \int t dN \Big/ \int dN \\ &= \int_0^\infty t e^{-3Ht}\, dt \Big/ \int_0^\infty e^{-3Ht}\, dt \\ &= \frac{1}{3H} . \end{aligned}$$

This mean age is independent of the epoch of observation, as demanded by the Perfect Cosmological Principle.

Because the Universe is assumed to be homogeneous and isotropic the results of Robertson and Walker show that the line element has to look like equation (2), and by the Perfect Cosmological Principle the curvature of space, $R\,a(t)$, must be constant, so the only possibility is $R = 0$. Since we know the geometry of space the energy flux f from a galaxy of known luminosity $\mathfrak{L}$ at high redshift is directly fixed, with no more free parameters. What is more we are assured in this model that galaxies at high redshift, which we observe as they were in the distant past, are on the average like the galaxies we see near us. Thus one can state unambiguously the expected deviation of the galaxy counts, $N(>f)$, from the naive law (18) knowing the properties of galaxies reasonably close to us.

By 1948 Hubble's value for H had been re-examined and found reasonable. Bondi, Gold and Hoyle could point to the discrepancy between the maximum age in the Big Bang model, $H^{-1} \sim 2 \times 10^9$ y if $\Lambda = 0$, and the radioactive decay ages, 3 to 6×10^9 y, which perhaps showed that the Lemaître model is inadequate. They could also hope for a clear decision from the galaxy counts and the other cosmological tests of the sort described in Chapter VI, because the Steady State model gave unambiguous predictions (to the extent that one knows the local parameters). As it turned out H^{-1} has since been revised upward by a factor of 5 at least, and it is still unclear whether there is a time scale problem in the Lemaître model (Chapter III). The galaxy count test has been transformed to radio source counts, and the data on the face of it go against the Steady-State model,[42] although Hoyle has argued that this may be deceptive.[43] The major observational objection to the Steady State model is the possible discovery of the Primeval Fireball radiation (Chapter V). If this radiation proves to have a thermal blackbody spectrum it will be direct and tangible evidence that the Universe passed through a dense and hot phase where thermal relaxation would have been possible. This would be in direct conflict with the Perfect Cosmological Principle as it appears to have been originally conceived. Of course one can get around this. In a model considered by Hoyle and Narlikar the Principle is satisfied only in the very long time average, continuous creation coming in bursts between which the Universe expands and evolves much as the Lemaître model.[44] If the burst of creation makes a dense and hot enough state it could produce a thermal fireball.

Whatever the outcome of these tests the Steady State picture has had a beneficial effect on the subject. It was a stimulus to the development of the theory of nucleosynthesis in stars, for in the Steady State picture one could not fall back on element formation in a Big Bang. It was an important stimulus also to observational work, much of it designed to prove the cosmology wrong. The problem of the origin of the Universe is neatly dealt with in this picture as a continuous and ongoing process. By contrast we trace the Lemaître cosmology back to a singular state (where the equations and perhaps the Universe blow up) at a finite time in the past. This applies also to the original Lemaître model, for the initial Einstein model is indeed unstable, so it cannot have lasted infinitely long because there are inevitably thermal fluctuations given enough time. On the negative side the most common and severe objection to the Steady State picture is that it was devised from pure thought by hand to fit the one main observational fact, the law of general recession of the nebulae. This circumstance of course does not rule out the model, but when one recalls how adept theorists are at finding explanations to observations, whether the observations are valid or not, one may feel uneasy.

e) *Is the Universe Expanding?*

A tenet of almost every cosmology, whether evolving or Steady State or otherwise, is that widely separated galaxies are moving apart at the rate $v \cong H\ell$.[45] It is a worthwhile summary of the status of observational cosmology as described in varying detail in the following chapters to list here the evidence in support of this idea. The following points seem to be relevant.

(1) The apparent recession velocities based on the redshift in the spectral lines of galaxies show a direct linear connection with distances inferred from the apparent luminosity of the galaxy (and, in some closer galaxies, the distances inferred from angular size and brightness of nebulae within galaxies). The most precise test by Sandage verifies the relation $v \propto \ell$ in clusters of galaxies with a scatter of only 15 percent in apparent recession velocity over the range of $\sim$ 1000 km sec^{-1} to $\sim$ 60,000 km sec^{-1} (Chapter II).

(2) The apparent radial velocities inferred from the frequency shift of the 21 cm radio line from atomic hydrogen agree with the optical values, as would be expected from the Doppler effect.[46] The two measures have been compared over the apparent velocity range $-$ 300 km sec^{-1} to 4000 km sec^{-1} (the minus sign meaning blue shift, motion toward us) and there is no appreciable evidence for systematic deviation over this range.

(3) Hubble's law, $v = H\ell$, is the functional form expected if the Universe is expanding and if the Universe is homogeneous and isotropic, as is suggested to modest accuracy by the galaxy counts and very precisely by the isotropy of the radiation background (cf. Chapter II).

(4) Hubble's law of the general recession is a consequence of the relativistic cosmologies people were considering in the 1920's. While there was not an *a priori* prediction of the effect the very natural way in which theory and observation meshed was and is considered strong evidence in favor of the expansion hypothesis. In the case of the Steady State picture the model came well after the discovery of the effect, but it was argued that it should have been expected ahead of time, for if the Universe were not expanding the Perfect Cosmological Principle would guarantee that the perpetually radiating galaxies would fill space with blackbody radiation at the highest temperature at which energy can be gained by nuclear reactions or any other process. A convenient sink for radiation and for dead stars and galaxies is the expansion of the Universe.[47]

(5) The numerical value of the constant of proportionality in Hubble's law defines a cosmic time scale H^{-1}, and in the evolving expanding cosmologies (aside from some special choices of parameters) one would look for approximate agreement of H^{-1} with maximum star ages and radioactive decay ages of the elements. As discussed above and in Chapter III there is agreement within a factor of 2 or better. In the Steady State cosmology the mean age of the galaxies in a large volume of space is $H^{-1}/3$, so one might again look for an order of magnitude coincidence, although the age of any chosen galaxy could be very much larger or smaller than H^{-1}.

(6) The Universe may contain a Primeval Fireball, blackbody radiation left over from a time when the Universe was dense and hot (Chapter V). If this is substantiated by further measurement it will be direct evidence that the Universe really is expanding and growing less dense, in agreement with the Lemaître cosmology (but not the original Steady State model).

(7) A direct test of the expansion hypothesis is to look for relativistic corrections to observed quantities like apparent magnitude or angular size

at high redshift, where the apparent recession velocity is close to the velocity of light. Hubble first attacked this problem, in his counts of galaxies as a function of apparent magnitude, and in recent years this problem has been intensively pursued by Sandage and others. It is not straightforward even in principle because there are a number of effects operating, so for a test one ultimately needs enough lines of attack to over-determine the parameters. It appears, however, that the available data may be adequate to rule out the most naive "tired light" alternative to the expansion postulate (Chapter VI). The radio source counts as functions of energy flux show a strong deviation from the naive law, but until the sources are more completely identified and understood it will be hard to separate expansion effects from other effects, like intrinsic evolution of the sources.

(8) Perhaps the most serious potential problem for the expansion hypothesis is the apparent tendency of quasar absorption line redshifts to cluster around preferred values.[48] If substantiated this would mean either that large redshifts can be produced at least in some cases by effects other than the cosmological expansion, or that the Universe shows a striking and unexpected tendency to place absorbing clouds at preferred redshifts (preferred distances).

It should be apparent from this list that the expansion postulate, while strong, does not enjoy the overwhelming weight of evidence of, say, the quantum principle. What is needed is a more tightly woven web of interconnected results, and one of the major goals of cosmology still is to build this web. Following the almost exclusive practice in cosmology the expansion of the Universe will be adopted below as a convenient reasonable and fertile working hypothesis, but it is well to bear in mind that it is still an hypothesis.

REFERENCES

1. E. Hubble, *Realm of the Nebulae* (Yale) 1936 (Dover Publication, 1958).
2. V. M. Slipher, Lowell (Flagstaff) *Obs. Bull.* 58, 1914.

3. A. S. Eddington, *Mathematical Theory of Relativity* (Cambridge) 2nd ed., 1924, p. 162; this was discussed also by K. Lundmark, *M. N.* 84, 747, 1923.

4. W. deSitter, *M. N.* 78, 3, 1917.

5. A. Einstein, *S.-B. Preuss. Akad. Wiss*, 142, 1917; English tr. in *The Principle of Relativity* (Dover Publication).

6. J. D. North, *Measure of the Universe* (Oxford) 1965.

7. For a good description of the Kapetyn universe and other early speculation on the nature of the Milky Way Galaxy see O. Struve and V. Zebergs, *Astronomy of the 20th Century* (Macmillan) 1962, Chap. 19.

8. H. Shapley, *Ap. J.* 48, 89, 1918. Shapley gives a very readable informal account of his work in *Through Rugged Ways to the Stars* (Charles Scribners, New York, 1969).

9. A. van Maanen, *Ap. J.* 57, 264, 1923.

10. E. Hubble, *Ap. J.* 62, 409, 1925.

11. H. Shapley, *Harvard Bull.* 796, 1923.

12. E. Hubble, *Proc. Nat. Acad. Sci.* 15, 168, 1929.

13. H. Weyl, *Phys. Z.* 24, 230, 1923.

14. H. P. Robertson, *Phil. Mag.* 5, 835, 1928.

15. A. Friedman, *Z. f. Physik* 10, 377, 1922.

16. G. Lemaître, *Ann. Soc. Sci. Bruxelles* 47A, 49, 1927; English translation in *M. N.* 91, 483, 1931.

17. A. Einstein, *S.-B. Preuss. Akad. Wiss.*, 235, 1931.

18. H. P. Robertson, *Proc. Nat. Acad. Sci.* 15, 822, 1929.

19. E. A. Milne, *Relativity, Gravitation and World Structure* (Oxford) 1935, p. 73.

20. E. A. Milne and W. H. McCrea, *Quart. J. Math.*, Oxford Ser. 5, 64 and 73, 1934.

21. C. G. Callan, R. H. Dicke and P. J. E. Peebles, *Amer. J. Phys.* 33, 105, 1965.

22. R. C. Tolman, *Relativity, Thermodynamics and Cosmology* (Oxford) 1934, p. 252.

23. E. T. Whittaker, *Proc. Roy. Soc.* A149, 384, 1955; ref. 22, Sec. 93; W. H. McCrea, *Proc. Roy. Soc.* A206, 562, 1951.
24. H. Shapley and A. Ames, *Ann. Harvard College Obs.* 88, 41, 1932; H. Shapley, *Proc. Nat. Acad. Sci.* 19, 591, 1933.
25. E. Hubble, *Ap. J.* 79, 8, 1934; *Ap. J.* 84, 517, 1936.
26. A. Einstein and W. deSitter, *Proc. Nat. Acad. Sci.* 18, 213, 1932.
27. W. deSitter, *B. A. N.* 5, 157, 1930.
28. A. S. Eddington, *M. N.* 90, 668, 1930.
29. G. Lemaître, *The Primeval Atom* (D. Van Nostrand) 1950.
30. A. Holmes and R. W. Lawson, *Amer. J. Sci.* 13, 327, 1927.
31. For a description of this calculation see H. Jeffreys, *The Earth* (Cambridge) 2nd ed., 1928, Chap. 5.
32. F. W. Aston, *Nature* 123, 313, 1929; E. Rutherford, *Nature* 123, 313, 1929.
33. J. Jeans, *Nature* 122, 689, 1928; *Astronomy and Cosmogony* (Cambridge) 1928, Chap. XII.
34. A. S. Eddington, *M. N.* 84, 308, 1924.
35. A. S. Eddington, *Rotation of the Galaxy,* Halley Lecture, 1930 (Oxford); *The Expanding Universe* (Macmillan) 1933, Chap. III, Sec. IV.
36. W. deSitter, *M. N.* 93, 628, 1933.
37. This point is most recently emphasized by H. Alfvén, *Physics Today,* Feb., 1971 and O. Klein, *Science* 171, 339, 1971 (cf. also earlier references in these two papers).
38. G. C. McVittie, *Cosmological Theory* (Methuen) 1937, p. 67.
39. H. A. Bethe, *Phys. Rev.* 55, 103 and 434, 1939.
40. B. J. Bok, *M. N.* 106, 61, 1946.
41. H. Bondi and T. Gold, *M. N.* 108, 252, 1948; F. Hoyle, *M. N.* 108, 372, 1948.
42. A recent review is given by M. Ryle, *Ann. Rev. Astron. Astrophys.* 6, 249, 1968.
43. F. Hoyle, *Proc. Roy. Soc.* A308, 1, 1968.
44. F. Hoyle and J. V. Narlikar, *Proc. Roy. Soc.* A290, 162, 1966.

45. F. Zwicky has most persistently and critically discussed the expansion hypothesis – see for example *Morphological Astronomy* (Springer), 1957.

46. M. S. Roberts, IAU Symposium #44, to be published; W. K. Ford, V. Rubin and M. S. Roberts, *A. J.* 76, 22, 1971.

47. H. Bondi, *Cosmology* (Cambridge) 2nd Ed., 1960, Chap. III.

48. G. R. Burbidge and E. M. Burbidge, *Nature* 222, 735, 1969, and earlier references therein; for another discussion of the interpretation of redshifts see H. Arp, *Ap. J.* 148, 321, 1967.

II. THE HOMOGENEITY AND ISOTROPY OF THE UNIVERSE

A common assumption in cosmology is that the Universe is homogeneous and isotropic in the large scale, when minor irregularities like galaxies and clusters of galaxies are smoothed over. The major exceptions to this assumption have been the bounded world picture of the sort discussed by Klein and by Alfvén in which it is assumed that the galaxies about us are a finite metagalactic system expanding into more or less empty space (ref. I-37), and the hierarchical world picture, in which the observed sequence of structures ranging from star clusters to galaxies to pairs and groups of galaxies to great clusters of galaxies is assumed to continue to superclusters and so on to structures on the largest scale we can hope to see.[1] As will be described in this chapter the evidence is meagre but does suggest that the original assumption of homogeneity and isotropy has been remarkably successful.

a) *Clusters and Superclusters*

In 1932 Shapley and Ames published a catalogue of the brighter galaxies (ref. I-24). Their map of the positions of these galaxies in the sky is convincing evidence that the galaxies around us are far from uniformly distributed.[2] Many are concentrated in a small patch in the sky, the Virgo cluster of galaxies, and many more are found along a swath running away from this cluster. Apparently if the Universe is homogeneous the necessary averaging length must be greater than the distance to the Virgo cluster, about 10 Mpc.

There are many other clusters of galaxies comparable in size to the Virgo cluster, among them the rich compact clusters, roughly similar to the Virgo cluster in total luminosity but more compact and more nearly centrally symmetric. Abell has made a catalogue of some 1500 of these rich clusters.[3]

He points out that they make prominent landmarks by which one can hope to map out the large scale distribution of matter in the Universe. Abell has suggested that the rich clusters show some tendency to be grouped in second-order clusters, or superclusters, and indeed he can point to some specific apparent examples of the effect. The suggested superclustering scale is ~ 50 Mpc. However, an objective statistical analysis of the catalogue data fails to reveal a statistically significant effect save in one of Abell's distance classes in one hemisphere.[4] This analysis does not rule out the supercluster effect, only limits it to 10 percent of the rich clusters concentrated in the assumed superclusters of 10 members each. To this sort of accuracy it appears that large scale clustering is not apparent in Abell's data, which reaches to redshift $Z \sim 0.2$, corresponding to a distance of perhaps 1000 Mpc. This goes against the intuitive feeling of some astronomers that the most distant clusters are distributed in a decidedly irregular fashion. While it may be that the distribution is irregular, the problem has been to find objective evidence of it in the face of limited statistics and the inevitable systematic errors.

b) *Galaxy Counts*

As was described in Chapter I, one can test for homogeneity by counting galaxies as a function of limiting magnitude. Hubble made the deepest optical survey of this kind (ref. I-25). To decide how far out Hubble was surveying one needs the luminosity function for galaxies, that is, the relative numbers of galaxies of different absolute luminosities. The following is a convenient and reasonable approximation to the mean number of galaxies per unit volume brighter than absolute magnitude M:[5]

$$\begin{aligned} n(<M) &= A\,10^{\alpha M}, \quad \alpha = 0.75, \; M < M^*, \\ n(<M) &= B\,10^{\beta M}, \quad \beta = 0.25, \; M > M^*, \end{aligned} \tag{1}$$

where the constants A, B and M^* satisfy the continuity condition

$$A\,10^{\alpha M^*} = B\,10^{\beta M^*}.$$

The magnitude scale used in this equation is explained in the Appendix.

It should be emphasized that this expression is intended only as a convenient analytic approximation for the purpose of deriving some numerical estimates. It is manifestly false in the limit of very dim galaxies ($M \to \infty$). The discontinuity in slope of equation (1) at M^* implies a discontinuity in value of the differential luminosity function, which may only be the result of approximating a smooth curve by two straight lines, or may possibly correspond to a real (but continuous) wiggle in the differential luminosity function near M^*.[6]

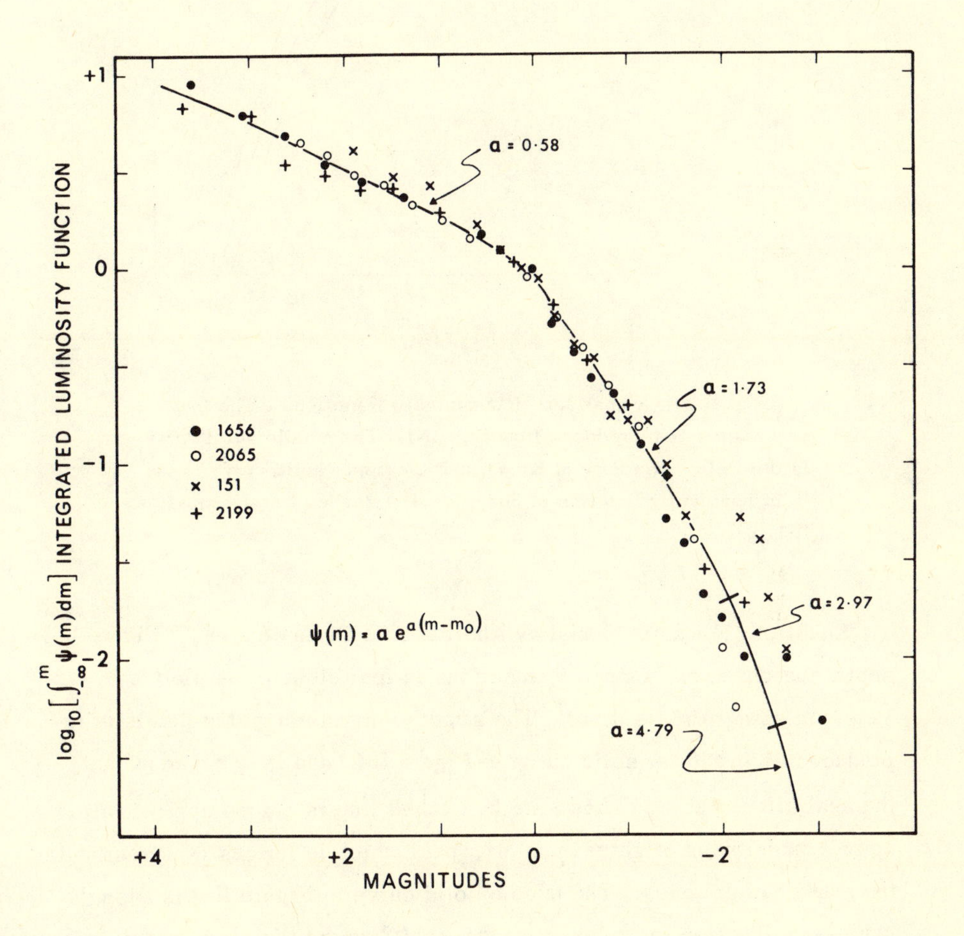

Fig. II-1. Luminosity Function for Galaxies in Rich Clusters.

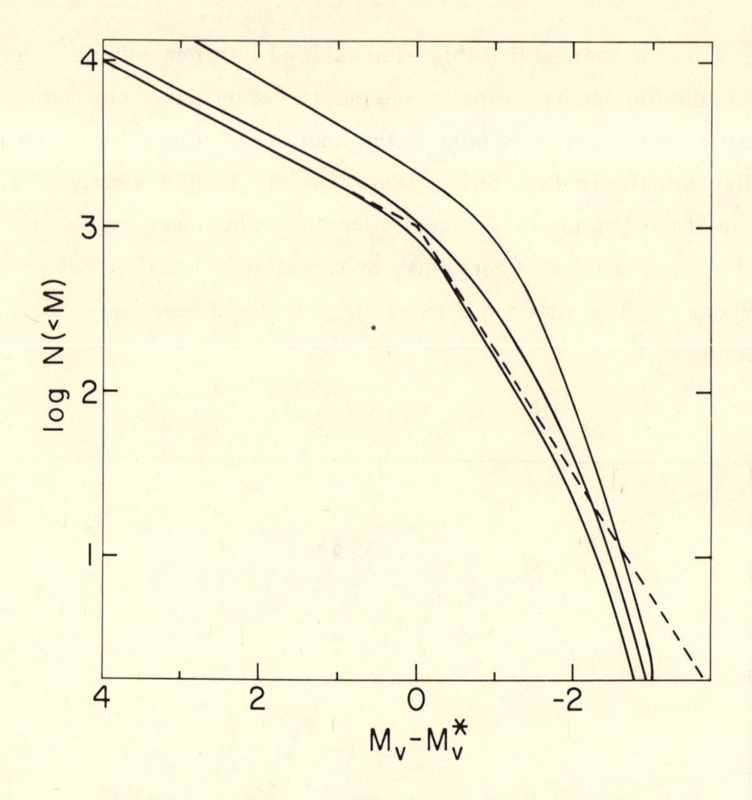

Fig. II-2. Comparison of Luminosity Functions. The lower solid curve is reproduced from Fig. II-1. The middle solid curve is the field luminosity of Kiang, and the upper solid curve is the field luminosity function of Shapiro. The dashed line is equation II-1.

Equation (1) was proposed by Abell for the Coma cluster.[5] Figure II-1 shows the observed luminosity functions in four clusters as plotted by Peterson using Abell's data.[7] The smooth curve through the points is reproduced as the lower solid curve in Figure II-2, and it is apparent that the analytic expression shown as the dashed line is a good approximation up to 2 magnitudes brighter than M^*, past which it seriously overestimates the number of galaxies. The middle solid curve in Figure II-2 is Kiang's luminosity function for galaxies in the field,[8] where the photographic magnitudes were converted to Abell's photovisual magnitudes by the expression

$M_{pg} - M_v = 0.9$, and of course the normalization of $n(< M)$ is freely adjustable. A more recent estimate of the field luminosity function by Shapiro is shown as the upper solid curve.[9] Again the agreement with equation (1) is fairly good, as it is for other surveys.[10] It was by no means obvious that the luminosity functions in the field and in compact clusters should be so similar, for the galaxies are systematically different, the abundant spiral galaxies in the field apparently being replaced by SO galaxies in compact clusters.[11] Whatever the explanation for this rough coincidence, it does appear that (1) is a reasonable basis for numerical estimates.

For the Coma cluster the apparent visual magnitude at the break in equation (1) is[5]

$$m_v{}^* = 14.7. \tag{2}$$

This is not corrected for absorption in the Galaxy. The correction might decrease $m_v{}^*$ by at most 0.2 magnitudes. The recession velocity of this cluster is

$$v = 6925 \text{ km sec}^{-1},$$

large enough that the distance might be estimated from Hubble's law (eq. I-1). Because Hubble's constant is so uncertain it is useful to write

$$H = h \times 100 \text{ km sec}^{-1} \text{ Mpc}^{-1}, \tag{3}$$

where h is thought to be between 0.5 and 1.0 (Chapter III). Then the distance to the cluster is

$$\ell = 69\, h^{-1} \text{ Mpc},$$

and, as described in the Appendix (eq. A-6), the apparent magnitude (2) translates to an absolute visual magnitude

$$M_v{}^* = -19.5 + 5 \log h. \tag{4}$$

Let us consider now the galaxy counts. The goal is to estimate how far Hubble was counting, neglecting refinements like expansion and curvature. There are $(dn/dM) \times dM$ galaxies per unit volume with absolute magnitude in the range of M to M + dM. These galaxies would be counted down to limiting apparent magnitude m if their distance were less than (eq. A-6)

$$\ell(M) = 10^{0.2(m-M)} \times 10 \text{ pc}. \tag{5}$$

Thus the number of galaxies per steradian brighter than apparent magnitude m is

$$\begin{aligned} \frac{dN}{d\Omega}(< m) &= \int_{-\infty}^{\infty} \frac{dn}{dM} dM \frac{\ell(M)^3}{3} \\ &= 10^{0.6m} \frac{(10pc)^3}{3} \int_{-\infty}^{\infty} dM \frac{dn}{dM} 10^{-0.6M}. \end{aligned} \tag{6}$$

This expression varies with limiting magnitude m as $10^{0.6m} \propto f^{-3/2}$, which is to be expected because it was assumed that the galaxies are distributed in a homogeneous fashion.

By equation (1) the integrand in (6) is a sort of bell-shaped function of M, peak at M^*, falling off toward the bright end $(M \to -\infty)$ as $10^{(\alpha-.6)M} \cong 10^{.15\,M}$, and falling off toward the dim end $(M \to \infty)$ as $10^{(\beta-.6)M} \cong 10^{-0.35\,M}$. That is, when galaxies are selected according to apparent magnitude the ℓ^3 factor strongly weighs the selection toward the intrinsically bright objects. For this reason it was some time before people realized that there are many more dim galaxies $(M > M^*)$ than bright per unit volume of space. The important point here, however, is that unless the luminosity function rises well above (1) at the dim end Hubble must have been counting mainly bright objects, $M \sim M^*$. Thus a reasonable estimate of the distance to which he was surveying is the distance ℓ^* at which an object with absolute magnitude M^* would have apparent magnitude equal to the limit of the survey. For a slightly better estimate one can note that equations (1), (5) and (6) imply that half the galaxies brighter than apparent magnitude m are at distances greater than

$$\begin{aligned} \ell_{1/2} &\cong \left[\frac{2\alpha(3-5\beta)}{3(\alpha-\beta)}\right]^{1/(5\alpha-3)} \ell(M^*) \\ &\cong 2.1\,\ell(M^*)\,. \end{aligned} \tag{7}$$

Hubble counted to limiting magnitude $m = 21.0$ on his magnitude scale. According to Holmberg's[12] calibration of the number-magnitude relation this would correspond to photographic magnitude 20.7, or visual magnitude about 19.8. Furthermore Hubble found that the dimmest galaxies were underabundant by a factor of slightly more than two, which may mean that he was overestimating the magnitude at the dimmest end by as much as 0.7 mag. This would suggest that his limiting magnitude m_V may be as low as 19.1. By equations (4), (5) and (7) the depth of Hubble's survey is then

$$\ell_{1/2} \cong 1100\ h^{-1}\ \text{Mpc},$$

and the corresponding redshift is

$$z_{1/2} = H\, d_{1/2}/c \sim 0.4. \tag{8}$$

The value of h drops out because M^* was calibrated from the redshift of the Coma cluster.

Hubble found a systematic discrepancy between the observed counts and the counts expected on a naive homogeneous static Euclidian model normalized to the Shapley-Ames counts of nearby galaxies, the observed counts being a factor of 2 or so low for the dimmest galaxies. People are now inclined to consider this a systematic measuring error. This does not invalidate the very important point that, to a factor of 2 or so accuracy, Hubble did not encounter an edge of the Universe, apparently out to the impressively large redshift $z \sim 0.4$, and he did not encounter a clustering effect on the characteristic scale ~ 1000 Mpc with density contrast greater than a factor of 2 or so.

c) *Hubble's Law*

The linear form of Hubble's law for $v << c$ is a necessary consequence of a homogeneous isotropic expanding world model. The converse does not necessarily follow of course but it is highly suggestive. The most precise test of Hubble's law is based on the brightest galaxies in rich clusters, which seem to have nearly constant absolute magnitude. If this is so then by the inverse square law and Hubble's law the measured energy flux f should vary inversely as the square of the redshift (assuming $z << 1$ so relativistic corrections may be ignored) which is a relation between measurable quantities. Sandage has found that the standard deviation of redshift z from the expected law $f \propto z^{-2}$ is about 15%, the scatter presumably resulting from some combination of measuring uncertainty, intrinsic spread in luminosity and, possibly, deviation from Hubble's law.[13] The relation is established to $z \cong 0.4$.

One way to represent possible deviations from Hubble's law corresponding to anisotropic expansion is to write the velocity of a galaxy as

$$v^{\alpha} = H^{\alpha\beta} r^{\beta},$$

where r^{β} represents the Cartesian coordinates of the galaxy, with our position at the origin.[14] As usual it will be understood that repeated indices are summed. This can be regarded as a first order expansion of the velocity field, the zero order term being discarded as a peculiar velocity. The "Hubble coefficient" may be expressed as

$$H^{\alpha\beta} = H\,\delta_{\alpha\beta} + \Theta_{\alpha\beta} + \sigma_{\alpha\beta},$$

where $\Theta_{\alpha\beta}$ is anti-symmetric, and represents rotation, and $\sigma_{\alpha\beta}$ is symmetric with zero trace, and represents shear. The observed quantity is the line of sight velocity,

$$\frac{r^{\alpha} v^{\alpha}}{r^2} = H + \sigma_{\alpha\beta}\,\gamma^{\alpha}\gamma^{\beta}, \tag{9}$$

where γ^{α} represents the direction cosines looking toward the galaxy. For the great clusters the standard deviation in the left hand side of (9) is about 15%, which means that the shear contribution on the right hand side cannot be larger than about 20%. Trendowski[15] computed the least square fit of (9) to the older Humason, Mayall and Sandage cluster redshift data.[16] He found that the components of $\sigma_{\alpha\beta}$ expressed in a coordinate system rotated to make $\sigma_{\alpha\beta}$ diagonal are

$$\frac{\sigma_{11}}{H} = 0.10$$

$$\frac{\sigma_{22}}{H} = 0.03$$

$$\frac{\sigma_{33}}{H} = -0.13$$

That is, the direct observational limit on shear is about 10 percent of expansion.

d) *The Radiation Background*

The final test is the isotropy of the microwave and X-ray background radiation.[17] Like Hubble's law this is an indirect test – to the extent that the Universe is homogeneous and isotropic the radiation background must be isotropic, but if the radiation is isotropic it may mean only that it has been scattered, or that we are at the center of a spherically symmetric Universe.

The discovery and possible interpretation of the microwave background is described in Chapter V. The measured radiation brightness in the range of 20 cm to 0.3 cm wavelength is fitted to good accuracy by a blackbody distribution at temperature $T_0 = 2.7$ K. There have been a number of experiments to measure the degree of isotropy of this radiation, the measurement generally being a scan along a narrow band of the sky. Some results are listed in Table II-1.

Table II-1.

ISOTROPY OF THE RADIATION BACKGROUND

Reference	Wavelength	Angular Resolution	$\delta i/i$
Partridge and Wilkinson[18]	3.2 cm	24-hour	$\lesssim 0.0008$
Partridge and Wilkinson[18]	3.2 cm	$15°$	$\lesssim 0.005$
Conklin and Bracewell[19]	2.8 cm	$10'$	$\lesssim 0.002$
Conklin[20]	3.75 cm	24-hour	$\cong 0.0006$
Boughn, Fram and Partridge[21]	0.86 cm	24-hour	$\lesssim 0.006$
Epstein[22]	0.34 cm	$12'$	$\lesssim 0.05$
Penzias, Schraml and Wilson[23]	0.35 cm	$2'$	$\lesssim 0.02$
Schwartz[24]	0.3-1.6 A	24-hour	$\lesssim 0.01$
Schwartz[24]	0.3-1.6 A	$20°$	$\lesssim 0.04$

The first seven entries are microwave measurements. In the last column δi may be considered a standard deviation on the energy flux in a solid angle fixed by the second column, and the denominator is the mean energy flux computed on the assumption that the background has a blackbody spectrum at temperature 2.7 K.

The last entry is the X-ray limit of Schwartz.[24] The significance of the X-ray isotropy for cosmology was first emphasized by Wolfe.[25]

The high isotropy of the microwave and X-ray radiation in principle could be due to scattering, but the required optical depth is rather severe. For example, let us suppose that the collection of galaxies around us is a finite cloud expanding into empty space, the expansion rate at the edges relative to the center being $v \sim 0.5\,c$ (corresponding to the maximum observed redshifts of galaxies). If the radiation originates within this system then either the system is spherically symmetric and we are right at the center to high precision, which seems absurd, or else the optical depth for scattering looking through the system must satisfy $e^{-\tau} \lesssim 0.0006$ at 4 cm wavelength. But radio sources associated with galaxies at high redshift show no evidence of such strong attenuation. A second possibility is that the radiation fills the space into which the system of galaxies is expanding. If intergalactic space is transparent it means that our velocity relative to the frame defined by the "extra metagalactic" radiation is less than 300 km/sec, which seems absurd when we recall that galaxies very similar to nearby ones are seen to be receding at 100,000 km sec^{-1}. If the radiation is strongly scattered by intergalactic matter the optical depth again must be high, $e^{-\tau} \lesssim 0.0006$.

For the isotropic X-ray background we do not have useful observational limits on the opacity of space, but it is very hard to see what could produce the high opacity needed in the bounded metagalaxy models. The most important scattering process for X-rays and microwave radiation is thought to be Thomson scattering, which would affect alike X-ray, optical and radio radiation. We know from the linearity of the redshift-magnitude relation that the optical depth to $z = 0.4$ in the visible is less than one magnitude, while the X-ray observations need 5 magnitudes.

It will be of considerable interest to know whether the X-ray background is resolved into structure at higher angular resolution. If so it will be strong evidence that the X-rays do come from localized objects like galaxies, that the radiation has not been strongly scattered, and that the source matter is homogeneously and isotropically distributed on the large scale.

e) *Summary*

It is amusing to note that the depth of space measured via Hubble's law to $z \sim 0.4$ is just comparable to Hubble's survey. Why was Hubble, working prior to 1936, using the smaller 100'' telescope to photograph red galaxies with blue sensitive plates, able to survey as far as the most modern work? Simply because Hubble was taking direct photographs, concentrating the light on a small spot, instead of dispersing it to get a spectrum. Why has Hubble's program not been repeated with modern techniques? Most likely because, being statistical, it consumed enormous amounts of telescope time which it would appear is just not available now. In view of the great power and fundamental significance of the test it is to be hoped that it will be repeated someday.

The homogeneity and isotropy assumption has been more successful than we had any right to expect from the highly ordered arrangement of observed matter in the nearest 10 Mpc. It still is possible to assume a hierarchical or bounded Universe provided (1) the density contrast in structures on a scale of 1000 Mpc is less than a factor of around 2; (2) the microwave radiation is exempt from the hierarchies; (3) the curvature of space is homogeneous and isotropic, unaffected by the hierarchies, so that the microwave radiation suffers the same redshift to an accuracy of 6 parts in 10^4 looking in different directions; (4) the X-ray background also is independent of the hierarchies, a "primeval" phenomenon not produced by matter.

One must bear in mind that each of the tests considered here is subject to uncertain interpretation, as is the rule in cosmology. For example, Hubble's survey may have been much less deep than we have estimated because there may be many more faint sources than are indicated by equation (1). With this reservation we conclude that the Universe on the face of it apparently is accurately homogeneous and isotropic on the scale $cH^{-1} \sim 3000$ Mpc.

REFERENCES

1. G. de Vaucouleurs, *Science* 167, 1203, 1970.
2. J. Oort, in Solvay Conference on *Structure and Evolution of the Universe*, 1958, p. 163.

3. G. Abell, *Ap. J. Suppl.* 3, 211, 1958.
4. J. T. Yu and P. J. E. Peebles, *Ap. J.* 158, 103, 1969.
5. G. Abell, *Problems of Extragalactic Research,* ed. G. C. McVittie, 1962, p. 213.
6. H. J. Rood, *Ap. J.* 158, 657, 1969.
7. B. A. Peterson, *Nature* 227, 54, 1970.
8. T. Kiang, *M. N.* 122, 263, 1961.
9. S. Shapiro, *A. J.* 76, 291, 1971.
10. S. van den Bergh, *Z. Ap.* 53, 219, 1961; M. A. Arakelyan and A. T. Kalloglyan, *Astron. Zh.* 46, 1215, 1969; Engl. tr.-*Soviet Astronomy-A.J.* 13, 953, 1953.
11. A description of clusters is given by G. Abell, *Ann. Rev. Astron. Astrophys.* 3, 1, 1965. The SO galaxies were described by A. R. Sandage in the *Hubble Atlas of Galaxies* (1961).
12. E. Holmberg, *Midd. Lund. Obs.,* Ser. II, No. 136, 1957.
13. A. R. Sandage, *Physics Today,* Feb. 1970 and earlier references therein.
14. J. Kristian and R. K. Sachs, *Ap. J.* 143, 379, 1966.
15. C. J. Trendowski, Senior Thesis, Princeton University, 1967.
16. M. L. Humason, N. U. Mayall and A. R. Sandage, *A. J.* 61, 97, 1956.
17. P. J. E. Peebles and D. T. Wilkinson, *Sci. American* 216, June, 1967.
18. R. B. Partridge and D. T. Wilkinson, *Phys. Rev. Letters* 18, 557, 1967.
19. E. K. Conklin and R. N. Bracewell, *Nature* 216, 777, 1967.
20. E. K. Conklin, *Nature* 222, 971, 1969.
21. S. P. Boughn, D. M. Fram and R. B. Partridge, *Ap. J.* 165, 439, 1971.
22. E. E. Epstein, *Ap.J.* 148, L157, 1967.
23. A. A. Penzias, J. Schraml and R. W. Wilson, *Ap.J.* 157, L49, 1969.
24. D. A. Schwartz, *Ap.J.* 162, 439, 1970.
25. A. Wolfe, *Ap.J.* 159, L61, 1970.

III. HUBBLE'S CONSTANT AND THE COSMIC TIME SCALE

The constant of proportionality in Hubble's law (eq. I-1) is important because it fixes the linear scale of the Universe, hence indirectly the time scale (eqs. I-21,22) and the critical mass density (eq. I-19). It is the first parameter in the basic cosmological equation (I-10). The purpose of this chapter is to give a brief review of the observational evidence on the value of the constant H, and to discuss how H^{-1} compares with the cosmic time scales provided by stellar evolution ages and the radioactive decay ages of the elements.

a) *Hubble's Constant*

To determine H one needs the distance of an object with redshift so large that the peculiar velocity is negligibly small compared to the cosmological redshift. The value of Hubble's constant is uncertain because the maximum distance which can be reliably determined is less than the minimum distance at which peculiar velocities are known to be unimportant.

Hubble's estimate $H \cong 500$ km sec^{-1} Mpc^{-1} was based on two main steps: (a) Cepheid variables fixed the distances within the Local Group of galaxies, for example, the distance to the Andromeda Nebula, M31; (b) the brightest stars calibrated within the local group were used as distance indicators in galaxies outside the Local Group. Both steps subsequently were recalibrated. In 1950 Baade discovered that the distance scale in the Local Group was in error by a factor of about 2 because of an error in the calibration of the Cepheid period-luminosity curve. The story of this discovery has been very well told by Baade.[1] Briefly put, it is now known, mainly through the work of Baade, that in the first approximation stars may be separated into two great classes: the population I, consisting of stars like the Sun with relatively high metal abundance, and apparently including all the stars that were formed in the disc of the Galaxy, and the population II, consisting of stars with relatively low metal abundance. The Pop II stars are old, apparently among the first generations in the Galaxy, and they form a sort of halo around the disc of the Galaxy. The globular star clusters that Shapley used to map out the scale of the Galaxy are members of the Pop II. There are Cepheid variable stars of a sort in both populations. In the Pop II the variables mostly have shorter period and lower luminosity, and they are called RR Lyrae variables or cluster variables (because they are found in globular star clusters). In his calibration of the Cepheid period – luminosity relation Shapley made the reasonable guess that the "Classical Cepheids" of the Pop I and the cluster-type variables fall along a universal period-luminosity curve. It is now known that the classical Cepheid period-luminosity relation implies a luminosity brighter by a factor of about 4 than the luminosity of an RR Lyrae variable. This built-in inconsistency in the period-luminosity relation was only discovered when Baade could "study the two kinds of Cepheids side by side, so to speak," in an extragalactic nebula.[1] This Baade did shortly after completion of the 200 inch telescope. With the discovery that there are two period-luminosity relations belonging to the two populations it was necessary to decide which if either of the relations satisfied the old calibration by statistical parallaxes. The guess was that the RR Lyrae were the more reliably determined because they have greater velocities, hence larger and more reliable proper motions, and because the RR Lyrae are found out of the disc, away from serious interstellar obscuration, while the classical Cepheids are found in the disc where the absorption correction is difficult. This would imply that the classical Cepheids are a factor of 4 brighter than previously thought, and, since these were the stars used to fix the scale of the Local Group, that the Local Group is a factor of 2 larger than previously thought. The distance to M31 is now put at 700 kpc, with good agreement among a number of distance indicators.[2] Hubble's original value was a factor 2.6 smaller. Because the extragalactic distance scale is based on the Local Group, H is accordingly reduced from Hubble's 1936 value $H = 526$[3] to $H \sim 200$ km sec^{-1} Mpc^{-1}.

The second re-calibration of H came with Sandage's discovery that what Hubble had identified in distant galaxies as bright stars were H II regions, knots of hot stars surrounded by a plasma ionized by the stars.[4] The H II regions were identified as such with the help of red-sensitive photographic plates that permitted identification of the strong red recombination line Hα from the plasma. The individual resolved bright stars apparently are dimmer than the H II regions by a factor of about 5. This implied that Hubble's value for H should be reduced by another factor of about 2.2. Sandage concluded that H may be somewhere between 50 and 100 km sec^{-1} Mpc^{-1}.

Tammann and Sandage have completed an important step in fixing the distance scale outside the Local Group with the isolation and measurement of Cepheid variables in the galaxy NGC 2403, which may be a member of the M81 group.[5] This group of galaxies is roughly comparable in size to the Local Group. The dominant member is the spiral galaxy M81, the most famous member the exploding galaxy M82. The distance to NGC 2403 appears to be reliably fixed at $\ell = 3.25 \pm .20$ Mpc. The straight mean of the radial velocities (corrected for rotation of the Galaxy) of all identified group members is $v = 210$ km sec^{-1}, whence $v/\ell = 65$ km sec^{-1} Mpc^{-1}. Tammann and Sandage very properly emphasize that this ratio may be quite different from H mainly because of the uncertain velocity, it being necessary to know (1) our motion relative to the center of mass of the Local Group, (2) the motion of the center of mass of the M81 Group relative to us, (3) the peculiar velocities of the two centers of mass, (4) whether or not NGC 2403 is a member of the M81 Group. The problem in getting at the first two parts is that we think we understand the dynamics of the M81 Group at best only very poorly, and of the Local Group not at all (cf. IV-c-ii,iii). Eventually one may be able to ignore the Local Group by using the direct measurement of our peculiar velocity relative to the microwave background (cf. V-e-i), and it may then be possible to get at any asymmetry in the local expansion rate from the observed motions and distances of nearby groups in several parts of the sky. For the present, however, it seems clear that to get H we need another indicator to carry the distance measures further out.

Sandage has attempted to use globular clusters to fix the distance to the elliptical galaxy M87 in the Virgo cluster.[6] The luminosity function of some 2000 globular clusters around M87 was obtained by Racine, and Sandage proposed that the bright end of this distribution might be at the same absolute magnitude as the brightest globular clusters in M31 and the Galaxy, for which we have an apparently reliable distance scale. The resulting distance to M87 is 14.8 Mpc, assuming 0.25 mag. absorption in the Galaxy in the blue. Now one can proceed in several ways. The mean recession velocity for the list of galaxies in the Virgo cluster is 1136 km sec^{-1}, which would give

$$H = 1136/14.8 = 77 \text{ km sec}^{-1} \text{ Mpc}^{-1}. \quad (1)$$

Sandage argued that the brightest elliptical in the Virgo cluster agrees with the redshift-magnitude line for brightest galaxies in rich clusters, suggesting (a) that the brightest elliptical in Virgo is one of the "standard candles" and (b) that the recession velocity 1136 km sec^{-1} is not seriously in error because of peculiar motions.

Sandage and van den Bergh[7] have emphasized that the distance estimate in (1) may be in error because the proposed comparison of globular clusters around the Local Group and M87 may not be valid. Probably this question can be settled only when we have an independent precise measure of the distance to the Virgo cluster.

Abell has argued that there may be a second difficulty. The Virgo cluster is less compact and less regular than the rich clusters typically included in the redshift-magnitude diagram, so it may be unreasonable to assume that the brightest elliptical in the Virgo cluster is as bright as the "standard candles." However, the Virgo cluster still might be used as an intermediate link. By matching the shapes of luminosity functions for elliptical galaxies in clusters, rather than the brightest end point, Abell and Eastmond concluded that the Coma cluster is further away than the Virgo cluster by the factor 8.7, the Corona Borealis cluster further away than Virgo by the factor 27.5.[8] These two clusters are of the rich compact sort. Using Sandage's distance to Virgo, Abell and Eastmond could find the distances to these two clusters, 130 Mpc and 410 Mpc respectively, and from the recession velocities of the two clusters, 6866 km/sec and 21600 km sec^{-1}, they had two estimates of Hubble's constant, both of which came to

$$H = 53 \text{ km sec}^{-1} \text{ Mpc}^{-1}. \tag{2}$$

If the Abell-Eastmond value for H is correct it will mean that the brightest elliptical in the Virgo cluster is dimmer than the standard in rich clusters by a factor of 2, and that this galaxy fell close to the redshift-magnitude line for rich clusters through an accidental cancelling error of 40% in the redshift. An error this large in the redshift is perhaps not improbable. The mean redshift for all galaxies identified with the Virgo cluster is 1136 km sec^{-1}, but de Vaucouleurs has noted that there may be an elliptical cloud separate from the spirals, the mean redshift for the ellipticals being 950 ± 104 km sec^{-1}, where the indicated error is the standard deviation in the mean.[9] With Sandage's distance of 14.8 Mpc and the Abell-Eastmond H, the cosmological redshift would be $H\ell = 780$ km sec^{-1}, smaller than the E cloud mean by 1.6 standard deviations. On this interpretation the discrepancy between expected and observed redshift would not be statistically significant.

Van den Bergh has repeatedly made the very good point that it may be wise not to rely on any one distance indicator, but to use instead the mean result of the greatest possible variety of distance indicators. This should cancel errors to some extent, and it may even reveal inconsistencies. Van den Bergh concludes from a recent survey that the best mean is[7]

$$H \cong 95 \pm 15 \text{ km sec}^{-1} \text{ Mpc}^{-1}. \tag{3}$$

Within the very substantial random errors of the problem this is consistent with (1), but in serious disagreement with (2).

Each one of the above estimates for H is provisional and uncertain. The best we can conclude is that, in agreement with Sandage's 1958 assessment, H probably is between 50 and 100 km sec^{-1} Mpc^{-1}. In the convention of equation II-(3) this means $0.5 \lesssim h \lesssim 1$. A useful conversion factor is

$$H^{-1} = 0.98 \times 10^{10}\, h^{-1} \text{ years}. \tag{4}$$

b) *Stellar Evolution Ages*

Stellar evolution ages are derived from the attempt to fit observed physical properties of a star, like mass, luminosity, and surface temperature, to a numerical model for the evolution of the star. The main attention has been directed to star clusters, where it is assumed that all members of a cluster formed at the same time, and that the members are distinguished one from the other by only one parameter, the star mass. The lower mass stars evolve only very slowly so they look now much as they did when first formed. The more massive stars have had time to burn out the hydrogen to form a central helium core. When this happens the luminosity and radius increase, and the star leaves the main sequence to become, for a short time, a red giant. A convenient rough order of magnitude is that the star leaves the initial main sequence when it has burned about 10% of its hydrogen to helium. This means that the lifetime on the main sequence is

$$t_e \sim 1.0 \times 10^{10} \text{ y } \mathfrak{M}/\mathfrak{L},$$

where the stellar mass and luminosity are expressed in units of the solar mass and luminosity. Aside from this result it is difficult if not impossible to explain the situation by the approach of order-of-magnitude estimates favored in these notes. Therefore the following is intended only as a brief guide to the climate of opinion on this very important subject.

The situation up to 1963 has been described in detail by Sears and Brownlee.[10] By that time the computed evolution ages of globular star clusters had swung from the original result of Sandage and Schwarzschild,[11] $t_e \sim 3 \times 10^9$ y, to a high of 25×10^9 y a decade later.[12] This latter result was based on the assumption that the globular cluster stars were formed with negligible initial helium. An increased helium abundance increases the luminosity of a star with given mass because the greater molecular weight means that the central temperature must be greater to support the star, this means a greater temperature gradient, hence greater heat flux. With increased luminosity the evolution time scale is reduced. Thus Sandage remarked that if the initial helium abundance were 24 percent by mass the evolutionary age of the star cluster would be reduced from 26 to 16 billion years.[12] The current tendency has been to argue that the initial helium was in fact even somewhat higher than this, and for this and other reasons the model ages have been reduced. Recently Sandage[13] has presented new color-magnitude diagrams for four globular star clusters, and he has fitted these results to evolutionary models of Iben and Rood and of Demarque, Mengel and Aizenman. The adopted helium abundances were in the range of 30 to 35 percent by mass, the derived evolution ages clustering around 11×10^9 y. The evolution ages would be substantially larger if the initial helium were lower. Unfortunately, as described in Chapter VIII, there is still considerable divergence of opinion on the initial helium problem.

c) *Age of the Elements*

An important and reliable datum is the radioactive decay age of the Earth and meteorites. The present accepted value was first presented by Patterson,[14] who used the uranium decay rates. Patterson assumed that the Earth and each of the meteorites formed as separate isolated and closed systems all at the same time, that the original lead in each meteorite and in the Earth had the same isotopic composition, and that the uranium isotopic abundance ratios are the same in each sample. It will be recalled that U^{235} decays to Pb^{207}, U^{238} decays to Pb^{206}, and that the isotope Pb^{204} has no long-lived progenitors. Thus the measured lead and uranium isotopic abundances in a meteorite satisfy

$$R_1 \equiv \frac{Pb^{207}}{Pb^{204}} = \frac{Pb^{207}i}{Pb^{204}} + \frac{U^{235}}{Pb^{204}} (e^{\lambda_1 t_m} - 1),$$

$$R_2 \equiv \frac{Pb^{206}}{Pb^{204}} = \frac{Pb^{206}i}{Pb^{204}} + \frac{U^{238}}{Pb^{204}} (e^{\lambda_2 t_m} - 1), \tag{5}$$

where the indicated ratios are supposed to be abundance ratios by number, the subscript i refers to the initial isotopic abundance, λ_1 and λ_2 are the decay rates of U^{235} and U^{238}, and t_m is the age of the meteorite. If the ratios R_1 and R_2 are measured in two meteorites, a and b, we can by the assumptions eliminate the initial lead abundance ratio by subtracting the corresponding values of R_1 and of R_2, to get

$$\frac{R_1(a) - R_1(b)}{R_2(a) - R_2(b)} = k\,\frac{(e^{\lambda_1 t_m}-1)}{(e^{\lambda_2 t_m}-1)}, \tag{6}$$

where k is the present isotopic abundance ratio,

$$k = \frac{U^{235}}{U^{238}}.$$

In equation (6) the only unknown is t_m. Patterson showed that the age

$$t_m = 4.55 \times 10^9 \text{ y} \tag{7}$$

very precisely satisfies the meteorite data and the Earth lead isotopic abundance ratios found in oceanic sediments. Meteorite ages based on the decays $Rb^{87} \to Sr^{87}$ and $K^{40} \to Ar^{40}$ now are in good agreement with (7).[15]

Equation (7) is a reliable lower bound to the age of the Galaxy, for it seems sure that the Sun and Solar System formed along with the rest of the disc Pop I in place in the disc. Unhappily we do not have an independent measure of the time interval between the formation of the Galaxy and the formation of the Solar System.

Rutherford had remarked that it is possible to estimate the age of uranium from the abundance ratio and relative decay rate of the isotopes U^{235} and U^{238} (Chap. I). This idea was taken up by Burbidge, Burbidge, Fowler and Hoyle in their classic synthesis of the theory of element formation in stars.[16] They estimated that the production ratio of U^{235} to U^{238} should be 1.64 by number (later modified to 1.65[17]), this being in first approximation the ratio of the numbers of progenitors that can reach the isotopes by decay rather than being lost by fission. Burbidge, *et. al.* computed the age of the Earth's uranium on two different assumptions: (a) that all the uranium was produced at one event, and (b) that the uranium was produced at a constant rate for some span of time ending with the isolation of the Solar System some 4.6×10^9 y ago. Following Dicke[18] it is convenient for purposes of discussion to generalize this to a one parameter series of models in which a fraction P_i of the net uranium production for the Solar System came in a very rapid burst at time $t = 0$, say, and the remaining fraction $1 - P_i$ was produced at a uniform rate from $t = 0$ to $t = t_s$, the time of isolation of the Solar System matter. Explicitly, the fractional abundances n_1, n_2 of U^{235} and U^{238} by number in the interstellar medium are assumed to satisfy, for $0 \leq t \leq t_s$,

$$\frac{dn_1}{dt} = A_1 (P_i t_s \delta(t) + 1 - P_i) - \lambda_1 n_1,$$
$$\frac{dn_2}{dt} = A_2 (P_i t_s \delta(t) + 1 - P_i) - \lambda_2 n_2, \tag{8}$$

where the constants A_1, A_2 satisfy the production ratio

$$A_1/A_2 = 1.65,$$

and the decay rates are the same as in equation (5). It should be emphasized that equation (8) represents only a schematic model intended to span in an approximate way the interesting possibilities. More detailed models of course can and have been constructed, but in view of the uncertainties it is doubtful at this point that they would be any more reliable than (8).

From the time t_s of isolation of the Solar System to the present, $t_o = t_s + t_m$ (eq. 7), the isotopes decay freely. The solution to (8) corrected for the free decay over the time t_m gives the present isotopic abundance ratio as

$$\frac{n_1}{n_2} = 1.65 \frac{\left[P_i t_s e^{-\lambda_1 t_s} + (1-P_i)(1-e^{-\lambda_1 t_s})\lambda_1^{-1}\right]}{\left[P_i t_s e^{-\lambda_2 t_s} + (1-P_i)(1-e^{-\lambda_2 t_s})\lambda_2^{-1}\right]} e^{-(\lambda_1-\lambda_2)t_m}. \tag{9}$$

The unknowns in this equation are P_i and t_s. It will be noted that P_i is the fractional production ratio for the Earth's uranium. Dicke's parameter P[18] is the fractional production from the burst for the *present* interstellar material, and the two are related by the equation

$$\frac{P_i t_s}{1-P_i} = \frac{P(t_s + t_m)}{1-P}.$$

The fractions P_i and P are given as functions of the total age $t_0 = t_s + t_m$ in Table III-1.

Table III-1

URANIUM MODEL AGES

t_0 (unit 10^9 y)	P_i	P
6.65	1	1
7.0	0.88	0.72
8.0	0.68	0.47
10.0	0.48	0.33
12.0	0.35	0.25
18.3	0	0

The range in model ages in the Table is almost a factor of 3. Fowler and Hoyle[17] discussed the possibility of narrowing this range by including the Th^{232} data. They found that the prompt production model ($P = 1$) gave discordant uranium and thorium ages, the discrepancy being such that their value of the Solar System abundance ratio Th^{232} to U^{238} is 3.8, the value required for consistency in the prompt production model 3.3. Fowler and Hoyle concluded that the longer time scale continuous production models may be indicated. Perhaps the most serious question here is whether the *a priori* computation could be expected to do better than the 15 percent discrepancy between computed and "observed" uranium-thorium abundance ratio.

Dicke[18,19] attempted to narrow the range of models by an appeal to the possible course of evolution of the Galaxy. In brief it is argued that the disc must have formed within a few times 10^8 y after the formation of the halo. Because there are no known disc population stars with low heavy element abundance it appears that most of the heavy elements must have been produced "promptly," during the collapse to the disc. The basic assumption here is that the Galaxy formed from a collapsing gas cloud, the halo Pop II stars forming during the initial collapse and the Pop I forming after the residual gas had settled to the equilibrium disc configuration. The time interval between formation of halo and disc is thought to be comparable to the free fall time. A semi-empirical argument given by Eggen, Lynden-Bell and Sandage is that the halo stars are observed to have nearly radial orbits, and this would not have been expected if the halo stars had formed from a gas cloud that was slowly settling, not freely falling.[20] A purely theoretical argument is that there is no known way by which a more or less spherical gas cloud with the mass and radius of the Galaxy ($10^{11}\mathfrak{M}_{\odot}$, 20 kpc radius) could support itself for more than a free fall time. If the cloud were supported by the gas pressure the gas would have to be hot, $\sim 3 \times 10^6$ K, it would soon collisionally ionize itself, and it would then cool by bremsstrahlung emission in somewhat less than a collapse time. If the cloud were held up by turbulence it could only be supported for the dissipation time of the turbulence, which is roughly the turn-over time for the largest eddy, again $\sim 10^8$ y. It appears therefore that star formation in the halo could have proceeded at a leisurely rate only if there were some unknown energy source to hold up the gas, or if the gas were added to the halo a small amount at a time from some much more dilute source. The tentative conclusion from this argument is that the disc formed promptly, and because there are no known disc population stars with low heavy element abundance, that the fraction P_i must be large, greater than 0.5 let us say, and the uranium age therefore less than 10×10^9 y.

d) *Summary*

The three age measures considered here, H^{-1}, the greatest stellar evolution ages, and the radioactive decay age of the elements, each are thought to be known to an accuracy of better than a factor of 1.5. This means that most people would be surprised if future estimates went more than a factor 1.5 larger or smaller than the central values now under discussion. Of course there is ample precedent for surprises in cosmology.

There is a possibility of a concordance of the ages at about 10×10^9 y. This is the Einstein-deSitter age if $H = 70$ km sec^{-1} Mpc^{-1}, about midway in the modern range of estimates of this parameter. It also agrees with some recent estimates of stellar evolution ages for globular cluster stars if these objects had high initial helium, in rough agreement with the helium production in the naive Big Bang model (Chapter VIII). It is also the uranium production age on a "moderate" prompt production picture, $P_i \lesssim 0.5$ in Table III-1. It should be clear, however, that one can imagine many other concordant schemes. Also, depending on the cosmology, one might look for a discordance. In the Steady State picture the age of the Galaxy bears no necessary relation to H^{-1}. Dicke[21] has emphasized for some time that stellar evolution ages may exceed both H^{-1} and the radioactive decay ages because the strength of the gravitational interaction may be decreasing with time, causing the stellar evolution rate to have been much greater in the past. This point was first made (as a restriction on the varying G assumption) by Teller.[22]

REFERENCES

1. W. Baade, *P.A.S.P.* 68, 5, 1956.
2. S. van den Bergh, *J. Royal Astron. Soc.* (Canada) 62, 1, 1968.
3. E. Hubble, *Ap. J.* 84, 270, 1936.
4. A. Sandage, *Ap. J.* 127, 513, 1958.
5. G. A. Tammann and A. Sandage, *Ap. J.* 151, 825, 1968.
6. A. Sandage, *Ap. J.* 152, L149, 1968.
7. S. van den Bergh, *Nature* 225, 503, 1970.
8. G. O. Abell, to be published in IAU Symposium #44, 1970; G. O. Abell and S. Eastmond, *A. J.* 73, S161, 1968.
9. G. de Vaucouleurs, *Ap. J. Suppl.* No. 56, 6, 213, 1961.
10. R. L. Sears and R. R. Brownlee, *Stars and Stellar Systems,* Vol. VIII, *Stellar Structure,* ed. L. H. Aller and D. B. McLaughlin, 1965, p. 575.
11. A. Sandage and M. Schwarzschild, *Ap. J.* 116, 463, 1952.
12. A. Sandage, *Ap. J.* 135, 349, 1962.
13. A. Sandage, *Ap. J.* 162, 841, 1970.
14. C. Patterson, *Geochim. et Cosmochim. Acta.* 10, 230, 1956.

15. W. H. Pinson, C. C. Schnetzler, E. Beiser, W. H. Fairbain and P. M. Hurley, *Geochim. et Cosmochim. Acta.* 29, 455, 1965; D. A. Papanastassiou and G. J. Wasserburg, *Earth and Planetary Space Science Letters* 5, 361, 1969.
16. E. M. Burbidge, G. R. Burbidge, W. A. Fowler and F. Hoyle, *Rev. Modern Phys.* 29, 547, 1957.
17. W. A. Fowler and F. Hoyle, *Ann. Phys.* (N. Y.) 10, 280, 1960.
18. R. H. Dicke, *Ap. J.* 155, 123, 1969.
19. R. H. Dicke, *Nature* 194, 329, 1962.
20. O. J. Eggen, D. Lynden-Bell and A. Sandage, *Ap. J.* 136, 748, 1962.
21. R. H. Dicke, *Rev. Modern Phys.* 34, 110, 1962.
22. E. Teller, *Phys. Rev.* 73, 801, 1948.

IV. THE MEAN MASS DENSITY OF THE UNIVERSE

The mean mass density of the Universe, ρ, is the second of the parameters in the fundamental cosmological equation (I-10). Discussion of ρ is particularly difficult and interesting because for every conceivable form of matter we have to find from observational and theoretical arguments an estimate of the amount or an upper limit. A convenient reference point in the discussion of ρ is the Einstein-deSitter density (eq. I-19), which in the notation of equation (II-3) is

$$\rho_c = 1.9 \times 10^{-29}\, h^2 \text{ g cm}^{-3}, \qquad n_c = 1.1 \times 10^{-5}\, h^2 \text{ protons cm}^{-3}. \tag{1}$$

It should be emphasized that this is only a reference point. There is no *a priori* theoretical prediction that the mean mass density in the Universe must amount to the critical value (1). Also, it should become very clear in this Chapter that we have no guarantee that all significant contributions to the mass density will be in forms readily detected by us. The best that can be done is to systematically hunt for tests for all reasonably conceivable forms of matter. The great hope is that as the estimates are improved and the list of unknown contributions is reduced the sum will converge to a value consistent with the independently determined values of all other terms in equation (I-10), or some other such theoretical equation.

a) *The Mean Mass Density Due to Galaxies*

In estimating the mean mass density due to galaxies the first step is to write

$$\rho(G) = L\mathfrak{M}/\mathfrak{L}, \tag{2}$$

where L is the total luminosity per unit volume due to galaxies, and $\mathfrak{M}/\mathfrak{L}$

is a suitable mean value for the ratio of mass to luminosity for galaxies. To get L Oort[1] assumed a shape for the galaxy luminosity function, but with unknown normalization, and then fixed the normalization from the known coefficient in the counts of galaxies as a function of limiting apparent magnitude (eq. II-6). The following calculation along these lines is based on Abell's luminosity function, equation II-1. The normalization constant is A. To fix it we substitute equation (II-1) into (II-6). The result is

$$\frac{dN(<m)}{d\Omega} = \frac{(10pc)^3}{3} A 10^{(\alpha-.6)M^* + .6m} \left(\frac{\alpha}{\alpha-.6} + \frac{\beta}{.6-\beta} \right).$$

The observed number of galaxies brighter than m is (ref. II-12)

$$\frac{dN(<m)}{d\Omega} = 1.43 \times 10^{-5} \times 10^{0.6m} \text{ ster}^{-1},$$

where m is a visual magnitude, and in converting to visual magnitudes it was assumed that $m_v - m_{pg} = 0.85$. On combining these equations, with $\alpha = 0.75$, $\beta = 0.25$, and using equation (II-4), we find

$$A 10^{\alpha M^*} = 0.015 \text{ h}^3 \text{ Mpc}^{-3}. \quad (3)$$

It will be recalled that M^* is not corrected for extinction in the Galaxy. If this correction were $\Delta m_v = 0.2$, (3) would be increased by 30 percent.

By equation (II-1) (3) is the mean number density of galaxies brighter than $M^* \cong -19.5$. Because the prominent massive galaxies have $M \lesssim M^*$ equation (3) says that the density of mean (prominent, significant) galaxies is in the range of one to three per 100 Mpc^3. There are of course many more small ones.

By equations (II-1), (II-4), (3), and (A-7), the luminosity per unit volume due to galaxies is

$$\begin{aligned} L &= \int_{-\infty}^{\infty} \frac{dn}{dM} dM \, \mathfrak{L}(M) \\ &= 82.4 \, A 10^{(\alpha-.4)M^*} \left[\frac{\alpha}{\alpha-.4} + \frac{\beta}{.4-\beta} \right] \mathfrak{L}_\odot \quad (4) \\ &= 3.0 \times 10^8 \, h \, \mathfrak{L}_\odot \text{ Mpc}^{-3}. \end{aligned}$$

This quantity is expressed relative to the solar luminosity, the comparison being made in the wavelength band of the visual magnitude system, $\lambda \sim 5500$ A. Oort, van den Bergh and Kiang have estimated the luminosity density L, and all are in reasonable agreement with (4) (ref. 1, II-8, II-10).

To get the mass density (4) must be multiplied by the mass-to-light ratio for galaxies. Spiral galaxies like our own system and M31 exhibit approximate rotational symmetry, the bright stars and gas moving in approximately circular orbits in the disc. Knowing the circular velocity $\Theta(r)$ from the Doppler shift in spectral lines across the disc one has the gravitational acceleration Θ^2/r, and by approximate inversion of Poisson's equation via an assumption about the thickness of the mass distribution normal to the disc one can find the mass density, hence the galaxy mass. The results for spirals typically are in the range[2]

$$\mathcal{M}/\mathcal{L} \sim (1-10)\,h\,\mathcal{M}_\odot/\mathcal{L}_\odot .$$

The mass scales directly as the radius of the galaxy, hence as h^{-1}. The luminosity scales directly as the square of the distance, hence as h^{-2}.

In elliptical galaxies typically there is little gas, so it is not easy to get at internal velocities, but here and for spirals Page[3] has successfully pursued the analysis of binary systems, pairs of galaxies apparently in tight orbit around each other. For each pair of galaxies one knows only (1) $v_\parallel$, the relative velocity along the line of sight (from the Doppler shifts) and (2) $r_\perp$, the projected separation of the galaxies; however, by considering a collection of binary systems with assumed random orientations of the orbits one can get at the mean mass-to-light ratio. The results for spiral galaxies are in substantial agreement with the rotation curve masses. For elliptical galaxies Page finds $\mathcal{M}/\mathcal{L} \sim 50$ h.[3]

For the purpose of discussion let us adopt as the mean

$$\mathcal{M}/\mathcal{L} \sim 20h\,\mathcal{M}_\odot/\mathcal{L}_\odot . \tag{5}$$

The product of (4) and (5) is

$$\rho(G) \sim 4 \times 10^{-31}\, h^2 \text{ g cm}^{-3}, \tag{6}$$

and the ratio of this value to the critical Einstein-deSitter density is

$$\rho(G)/\rho_c \sim 0.02\,, \tag{7}$$

independent of h.

The rather crude calculation described here can be improved somewhat by using a numerical form for the empirically determined luminosity function instead of the analytic approximation as we have done, and by computing separately the luminosity functions of different classes of galaxies and multiplying each by the best guess about $\mathfrak{M}/\mathfrak{L}$ for the class. The results are not greatly different from (7) (ref. II-9, 10).

Now the central question is, how reliable is (7)? The answer depends on the systematic errors, as described in the next two sections.

b) *The Luminosity Density and the Light of the Night Sky*

The first possible systematic error in (7) is that the luminosity density L of equation (4) takes account only of galaxies large enough to be detected and counted in the luminosity function (II-1). There is good reason to worry about this because we can detect galaxies only by virtue of two coincidences of the orders of magnitude. First, the surface brightness of a galaxy typically is comparable to the brightness of the night sky, the upper atmosphere emission plus zodiacal light. For example, the surface brightness of our Galaxy looking normal to the disc from our position in it is roughly one third as bright as the night sky. The beautiful photographs of galaxies, say in the *Hubble Atlas,* show features down to surface brightness ~ 10% of the sky, while it is possible to trace galaxies down to 1% of the foreground sky brightness. The second coincidence is that galaxies are large enough to be distinguished from stars. There are known extragalactic objects (the Zwicky compact galaxies[4]) with angular size right down to the limit of resolution, and of course the quasars go much smaller. Arp has illustrated these two coincidences in a very striking graphical way.[5] The worry is of course that these may not be coincidences, that we may be studying only that end of a much broader distribution of surface brightness and linear size that happens to reach our limits of detectability. Fortunately there is one fairly direct test for the integrated luminosity density L, including objects like the extended halos of galaxies, or dwarf galaxies, or intergalactic stars, that would not be picked up in the sky surveys. This is the extragalactic contribution to the light of the night sky.[6]

The extragalactic radiation background resulting from all the galaxies has been widely discussed as a cosmological test,[7] but it is important to make clear what we think we are testing. One question is, given the mean luminous emission per unit volume of space, what is the resulting background brightness? This has figured prominently in cosmological discussions, particularly of Olbers' paradox (ref. I-47). We are considering here a different question – do we know the mean luminous emission per unit volume of space?

In first approximation the contribution to the sky surface brightness (ergs cm^{-2} sec^{-1} $ster^{-1}$) by galaxies is the product of the luminosity per unit volume $L/4\pi$ with the distance we can see, $\sim cH^{-1}$,

$$i \sim \frac{cL}{4\pi H}, \tag{8}$$

and the goal is to see whether this expected brightness based on (4) agrees with the observed contribution to the light of the night sky.

For a more detailed computation we need a formula for the time variation of radiation brightness in an expanding Universe. Let $i(\nu, t)$ be the radiation brightness, defined as the energy flux per steradian and per unit frequency interval (ergs sec^{-1} cm^{-2} Hz^{-1} $steradian^{-1}$). As usual we suppose i is isotropic and homogeneous (independent of position). In the time interval δt each photon suffers the cosmological redshift (eq. I-15)

$$\delta\nu = -\nu(\dot{a}/a)\,\delta t, \tag{9}$$

and, since the Universe has expanded in radius by the fractional amount $(\dot{a}/a)\,\delta t$, the number density of photons evidently decreases by the fractional amount

$$\delta n_\gamma/n_\gamma = -3\dot{a}/a\,\delta t.$$

The radiation brightness suffers this same fractional decrease because the photon density is reduced, and in addition the photon energy $h\nu$ and photon bandwidth $d\nu$ both decrease by fractional amounts $(\dot{a}/a)\,\delta t$. The loss of photon energy reduces the brightness by a further fractional amount $(\dot{a}/a)\,\delta t$. On the other hand the brightness is computed per frequency interval, and the decrease in bandwidth increases the brightness by the same fractional amount, so the last two effects just cancel. We have then

$$i(t + \delta t, \nu - \nu\dot{a}/a\,\delta t) - i(t, \nu) = -3i\,\dot{a}/a\,\delta t,$$

which yields the differential equation

$$\frac{\partial i}{\partial t} - \frac{\nu}{a}\frac{da}{dt}\frac{\partial i}{\partial \nu} = -3i\,\frac{\dot{a}}{a}.$$

Using equation (9), this can also be written as

$$\frac{d}{dt} i(t, \nu(t)) = -3\frac{\dot{a}}{a}\, i(t, \nu(t)). \tag{10}$$

When there is a radiation source $j(\nu)$ (ergs cm^{-3} sec^{-1} Hz^{-1} $steradian^{-1}$) this equation is modified to

$$\frac{d}{dt}\, i(t, \nu(t)) = -3\,\frac{\dot{a}}{a}\, i(t, \nu(t)) + cj(t, \nu(t)). \tag{11}$$

When expansion may be neglected this is just the standard equation for radiative transfer in the absence of absorption. When expansion is important the time derivative means a total derivative, taking account of the time variation of frequency indicated by equation (9).

In the present problem $j_\nu = L(\nu)/4\pi$, where L is the mean luminosity density of galaxies, and by equation (11) the contribution to the brightness of the background due to the galaxies is

$$i(t, \nu) = \frac{c}{4\pi}\int_0^t dt' \left(\frac{a(t')}{a(t)}\right)^3 L(t', \nu(t')). \tag{12}$$

We have, from equation (4) and an estimate of the spectrum of the radiation from galaxies, the expected present value of the function $L(t, \nu)$. However, to evaluate the integral we also need to know how L varies with time, that is, how the galaxies are evolving, and we need to fix on a cosmological model to determine the time variation of $a(t)$. When the integral (12) is evaluated for present frequency of observation in the far infrared ($\lambda \sim 3 - 10\mu$) the value of the integral does depend strongly on the adopted model, but when the frequency of observation is in the visible ($\lambda \sim 5000$ A) the model dependence almost drops out. This fortunate circumstance arises because the luminosity $L(\nu)$ falls off on going into the blue and ultraviolet. If ν now is in the visible, then on carrying the integral (12) back in time we soon find that $\nu(t')$ has shifted into the blue, and $L(t', \nu(t'))$ has become negligible. Thus we are only called upon to evaluate L at recent epochs, so we do not have to extrapolate very far back from present conditions.

Following this line of thought, we can evaluate the expected optical background by neglecting in the integral in (12) the relatively slow time variation of $a(t')^3$ and of galaxy luminosity. In this approximation we have, by (9),

$$
\begin{aligned}
i(\nu) &\cong \frac{c}{4\pi} \int_0^t L(\nu(t'))\,dt' \\
&= \frac{c}{4\pi} \int_\nu^\infty \frac{L(\nu')\,d\nu'}{\nu'(\dot{a}/a)} \qquad (13) \\
&\cong \frac{c}{4\pi H} \int_\nu^\infty \frac{L(\nu')\,d\nu'}{\nu'} .
\end{aligned}
$$

This may be compared with the simple estimate (8).

On using the measured spectra of giant galaxies[8] one finds that the integral in (13) carried over the wavelength range 3500 A to 5500 A is

$$\int_{3500}^{5500} L(\nu')\,d\nu'/\nu' = 0.20\, L(\nu = c/5500\text{ A}).$$

The contribution to the integral from $\lambda < 3500$ A would increase the coefficient in this expression from 0.20 to 0.24 if the luminosity per unit wavelength were constant shortward of 3500 A. The OAO observations suggest that L_λ may have a second strong peak shortward of 2000 A.[8] If so, it increases the expected sky brightness for a given photovisual luminosity per unit volume. To find a reasonable lower bound on the expected sky brightness let us ignore this possible effect, and take the coefficient to be 0.22. Then by equations (4) and (13) the expected integrated contribution to the background at 5500 A wavelength by galaxies is

$$\begin{aligned} \nu i_\nu &\cong 0.22\, c\nu\, L(\nu)/(4\pi H) \\ &= 5.0 \times 10^{-6} \text{ ergs cm}^{-2} \text{ sec}^{-1} \text{ ster}^{-1}, \\ &= 0.73\, S_{10}(V). \end{aligned} \tag{14}$$

The brightness is independent of h. We have expressed the result as the brightness per increment of the logarithm of frequency (or wavelength), $\nu i(\nu) = \lambda i(\lambda)$. In the last line the unit is the equivalent number of 10th magnitude (visual) stars per square degree (eq. A-5). This is a standard unit used by people who measure the light of the night sky.

Roach and Smith[9] give the following numbers for the contributions to the night sky near 5300 A wavelength:

Zodiacal Light $\sim 150\ S_{10}(V)$

Integrated Starlight $\sim 100\ S_{10}(V)$

Airglow continuum $\sim 50\ S_{10}(V)$.

Except for very isolated sites one must expect also a substantial contribution from electric lights. Roach and Smith measured the light of the night sky as a function of time and direction in the sky, and, working from the expected functional forms of each of these sources, attempted to isolate any possible extragalactic contribution. They set an upper limit $i \lesssim 5\ S_{10}(V)$ on the extragalactic light.[9] Lillie[9] has set an upper limit $i \lesssim 2\ S_{10}$ units at 4100 A. Apparently we can conclude that the luminosity density (4) is not in systematic error due to undetected objects by more than a factor of 5, say. We *know*, therefore, that we have not very seriously underestimated the possible contribution to L by dwarf galaxies, intergalactic stars, and the like.

c) $\mathfrak{M}/\mathfrak{L}$ *and the Stability Problem*

The second factor entering the mass density (7) is the mean mass-to-light ratio of galaxies. This quantity is suspect because when it is used to estimate the masses of groups or clusters of galaxies the result often appears to be unreasonable, as is described in the following examples. A useful collection of articles on the subject is in the *Astronomical Journal*, Vol. 66, December, 1961, where a number of earlier references will be found.

i) *The Coma Cluster of Galaxies*

This is the best studied of the rich compact clusters of galaxies. A description is given by Abell (ref. II-11). The mass of this system may be estimated by application of the virial theorem. Assuming the galaxies move like point masses interacting by gravitation alone the acceleration of the *ith* galaxy is

$$\frac{d^2 r_i}{dt^2} = \sum_{j \neq i} \frac{G m_j (r_j - r_i)}{|r_j - r_i|^3},$$

where m_j is the mass of the jth galaxy. On taking the scalar product of this equation with $m_i r_i$ and summing over i one finds

$$\begin{aligned} \frac{d^2}{dt^2} \sum \frac{m_i r_i^2}{2} - \sum m_i v_i^2 &= \sum_{j \neq i} G m_i m_j \frac{r_i \cdot (r_j - r_i)}{|r_j - r_i|^3} \\ &= \tfrac{1}{2} \sum_{j \neq i} G m_i m_j \frac{(r_i - r_j) \cdot (r_j - r_i)}{|r_j - r_i|^3} \qquad (15) \\ &= -\tfrac{1}{2} \sum_{j \neq i} \frac{G m_i m_j}{|r_j - r_i|}. \end{aligned}$$

The second equation follows by noting that the first double sum changes sign if the indices i and j are interchanged. The last term is just the gravitational potential energy of the system. If the system is in equilibrium, neither expanding nor collapsing, then the first term on the left hand side of the equation must have zero time average. It follows that, in the time average, the kinetic energy is one half of the magnitude of the potential energy,

$$\tfrac{1}{2} \sum m_i v_i^2 = \tfrac{1}{2}|U| \equiv \frac{G\mathfrak{M}^2}{2r_e}, \tag{16}$$

where $\mathfrak{M}$ is the cluster mass. The second equation defines the effective cluster radius r_e. Evidently one can use (16) to estimate the mass from the size and velocity dispersion in the object. This was first applied to clusters of galaxies by Zwicky and by Smith.[10]

In the Coma cluster the velocity dispersion and the spatial distribution of the galaxies appear to be in first approximation independent of galaxy mass, and the system is roughly spherically symmetric. This means that we can write the kinetic energy as

$$\begin{aligned} \tfrac{1}{2} \sum m_i v_i^2 &= \mathfrak{M} < v^2 > /2 \\ &= \frac{3}{2} \mathfrak{M} < v_{\parallel}^2 >, \end{aligned}$$

where $<v_{\parallel}^2>^{1/2}$ is the velocity dispersion observed along the line of sight. For the Coma cluster, the observed dispersion is

$$<v_{\parallel}^2>^{1/2} = 1020 \text{ km sec}^{-1}. \tag{17}$$

To get at the effective radius r_e Schwarzschild[11] proceeded as follows. Let $S(\delta)\,d\delta$ be the number of galaxies appearing in projection in the linear strip of width $d\delta$ placed a perpendicular distance δ from the cluster center. The function $S(\delta)$ can be directly obtained (after correction for background and foreground objects) from counts on a photographic plate. Assuming spherical symmetry S is related to the number density $n(r)$ by the formula

$$\begin{aligned} S(\delta) &= \int_0^{\eta(R)} 2\pi\, \eta d\eta\, n(r) \\ &= \int_{\delta}^{R} 2\pi\, r dr\, n(r), \end{aligned} \tag{18}$$

where R is the cluster radius, and $\eta = (r^2 - \delta^2)^{1/2}$. The mass of the cluster is

$$\mathfrak{M} = 2m \int_0^R S(\delta)\, d\delta\,, \tag{19}$$

where m is the mean mass of a galaxy. The potential energy is

$$|U| = 16\pi^2\, Gm^2 \int_0^R n(r)\, r dr \int_0^r n(r')\, r'^2 dr'. \tag{20}$$

By equation (18),

$$\frac{dS}{d\delta} = -2\pi\delta\, n(\delta)\,.$$

This equation can be substituted into equation (20), and the result integrated by parts twice, to find

$$\begin{aligned} |U| &= 4Gm^2 \int_0^R \frac{dS}{d\delta}\, d\delta \int_0^\delta \delta' d\delta' \frac{dS(\delta')}{d\delta'} \\ &= -\,4Gm^2 \int_0^R S(\delta)\, \frac{dS}{d\delta}\, \delta d\delta \\ &= 2Gm^2 \int_0^R S(\delta)^2\, d\delta\,. \end{aligned} \tag{21}$$

Equations (19) and (21) give

$$r_e \equiv G\mathfrak{M}^2/|U| = \frac{2\left(\displaystyle\int_0^R S d\delta\right)^2}{\displaystyle\int_0^R S^2 d\delta}. \tag{22}$$

This very pleasant equation permits one to obtain r_e by numerical integration of S, which in turn is directly obtained by counts of galaxies in strips. Schwarzschild used counts obtained by Zwicky to find $r_e = 1.48^\circ$. Omer, Page and Wilson have since repeated the counts in some detail.[12] The mean of the strip counts for the two 48-in. Schmidt plates counted by Wilson and by Omer and Page give

$$r_e = 1.69^\circ . \tag{23}$$

From the mean redshift for the galaxies in the cluster, 6925 km sec^{-1}, the distance is

$$R = 69\ h^{-1}\ \text{Mpc}, \tag{24}$$

and there is no reason to suspect that peculiar velocities cause a serious error. Equations (16), (17), (23) and (24) give

$$\mathfrak{M} = 1.5 \times 10^{15}\ h^{-1}\ \mathfrak{M}_\odot . \tag{25}$$

Abell estimates that there are 4000 rich clusters within $600\ h^{-1}$ Mpc (ref. II-11), which means a number density

$$n(RC) \sim 4 \times 10^{-6}\ h^3\ \text{Mpc}^{-3}.$$

(cf. eq. 3). If (25) is a typical mass of a rich cluster the mean mass density in this form becomes (ref. II-8)

$$\begin{aligned} \rho(RC) &\cong n(RC)\, \mathfrak{M} \\ &\sim 4 \times 10^{-31}\ h^2\ \text{g cm}^{-3}, \end{aligned}$$

just comparable to the estimated total mass density in galaxies (eq. 6)! The explanation is not that all galaxies are in rich clusters, but that if the mass estimate (25) is valid the rich clusters have unexpectedly large mass-to-light ratio.

Abell's estimate of the total luminosity of the Coma cluster is (ref. II-11)

$$\mathfrak{L} = 3.4 \times 10^{12}\ h^{-2}\ \mathfrak{L}_\odot ,$$

which by (25) implies a mass-to-light ratio

$$\mathfrak{M}/\mathfrak{L} \sim 400\ h\ \mathfrak{M}_\odot/\mathfrak{L}_\odot . \tag{26}$$

This is a factor of 20 larger than the assumed "typical value" (eq. 5). We can understand this in part because the Coma cluster contains elliptical and SO galaxies, for which the mass-to-light ratio, it is thought, might be in the range of 30-100. There remains a discrepancy of a factor of 5 to 10. We must consider now what this discrepancy might mean.

The virial theorem in a number of approximations has been applied to the Coma cluster, and there appears to be unanimous agreement that there is a formal discrepancy of a factor $\sim 3 - 10$. Zwicky has pointed out that the cluster surely does not have a well-defined radius, as is assumed in most analyses, including the above. The cluster just trails off into the general field. Also, it has been pointed out that, under the assumption of spherical symmetry, the velocity dispersion may be estimated from the dispersion in the observed line of sight velocities only if the sampling density is proportional to the surface number density of galaxies. If the galaxies in the clusters have radial orbits, and if the velocity dispersion is sampled only right at the center of the cluster, the indicated velocity dispersion will be too big by a factor of $\sqrt{3}$. Some assurance that this is not a serious problem is provided by an explicit numerical model for the formation of a cluster of galaxies.[13] The model reasonably well reproduces the observed surface density distribution in the cluster, and it does predict that the line of sight velocity dispersion should decrease with increasing distance from the cluster center. A fit of the model to the observed velocity dispersion across the face of the cluster, taking account of this expected variation of velocity dispersion, indicates that the cluster mass within the nominal cluster radius 100' arc is

$$\mathfrak{M}(r < 100') = 1.04 \times 10^{15}\, h^{-1}\, \mathfrak{M}_{\odot} \tag{27}$$

in reasonably good agreement with (25).

Despite the general agreement on a high $\mathfrak{M}/\mathfrak{L}$ ratio in the Coma cluster one should bear in mind that the different analyses are not statistically independent because they all perforce are based on much the same data. We urgently need comparable data on a number of clusters, to see if (26) is a statistically reproducible result.

If there is a mass anomaly what does it mean? In view of the smooth and regular appearance of the cluster probably most people would like to believe that it really is gravitationally bound and stable. There is in addition Zwicky's point that the cluster looks remarkably like an isothermal ideal gas sphere, which is strongly suggestive of an equilibrium situation (ref. I-45). The numerical model shows that 10^{10} y is time enough to establish a reasonable approximation to this equilibrium shape (but not long enough to establish energy equipartition).[13]

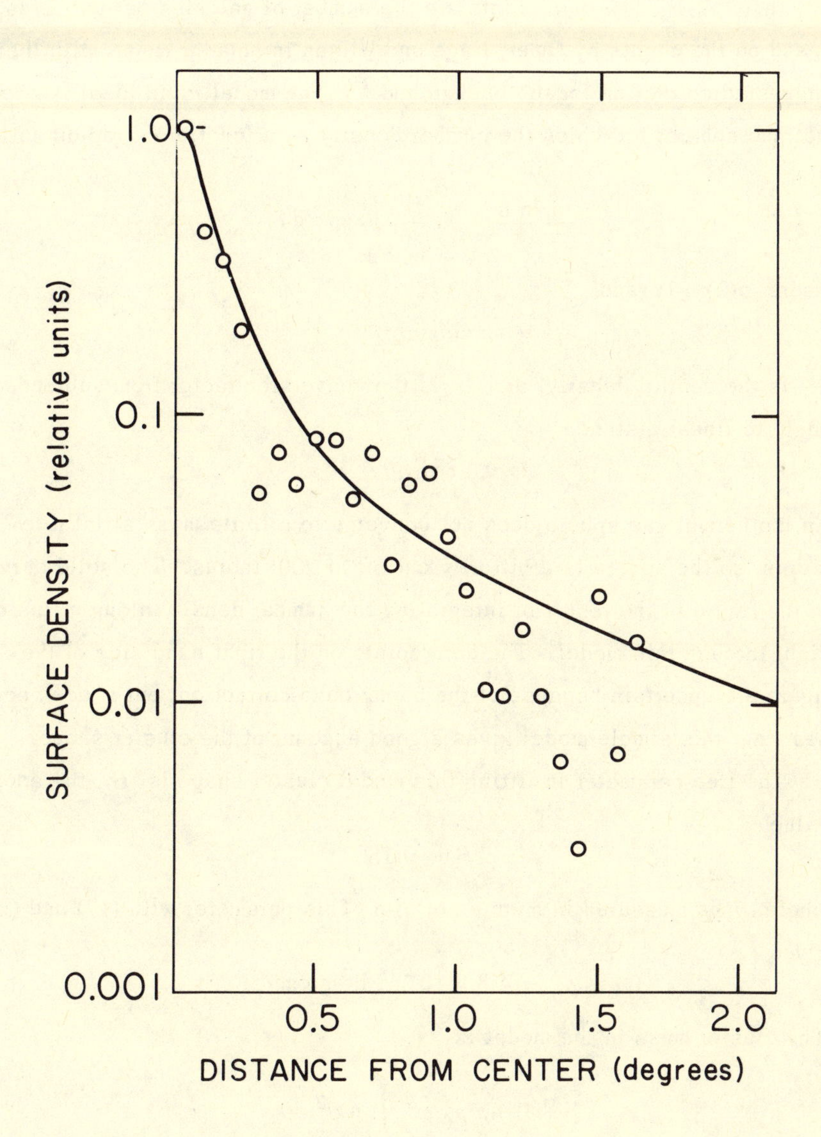

Fig. IV-1. Comparison of Coma Cluster Galaxy Counts with Ideal Isothermal Gas Sphere.

An application of the simple statistical equilibrium picture is illustrated in Figure IV-1. The data points are the number of galaxies per unit area, based on the counts by Omer, Page and Wilson in annular rings, with their suggested correction for the background.[12] The model is an ideal isothermal gas sphere, for which the number density as a function of radius satisfies

$$\frac{d \ln n}{d\theta} = -\frac{\beta}{\theta^2} \int_0^{\theta} n\theta^2 \, d\theta ,$$

where $n(0) \equiv 1$, and

$$\beta = 4\pi G\rho_0 \, a^2 \, / < v_{\parallel}{}^2 >^{1/2} .$$

ρ_0 is the central density, and a is the conversion factor from subtended angle to linear distance,

$$r = a\theta .$$

An isothermal gas sphere does not converge to a finite mass at infinite radius, so the sphere is arbitrarily cut off at 200' radius. The solid curve in the figure is the result of integrating the number density along a line of sight through this model. The data points on the right hand side of the figure are uncertain because of the background correction, but it does appear that this simple model gives a good account of the cluster shape.

The free parameter in fitting this model cluster shape is β, the adopted value

$$\beta = 0.15 ,$$

where θ is measured in minutes of arc. This parameter with (17) and (24) gives

$$\rho_0 = 4.8 \times 10^{-25} \, h^2 \, g \, cm^{-3} .$$

The cluster mass in the model is

$$\mathfrak{M} = 4\pi \, \rho_0 \, a^3 \int n \, \theta^2 \, d\theta .$$

This gives

$$\mathfrak{M}(<100') = 0.97 \times 10^{15} \, h^{-1} \, \mathfrak{M}_{\odot} ,$$

$$\mathfrak{M}(<200') = 1.86 \times 10^{15} \, h^{-1} \, \mathfrak{M}_{\odot} ,$$

the first number being the mass within the nominal cluster radius, the second the total mass to the artificially fixed outer boundary of the model. Again, the model mass agrees fairly well with (25).

The Coma Cluster and objects like it are very convenient for an investigation of a mass anomaly, always assuming there is one in such systems, because the parameters are fairly well fixed. As a first step let us consider another estimate of the central mass density, and an estimate of the surface density of the cluster.

When δ is very small (18) may be approximated as

$$S(0) - S(\delta) \cong \pi\, n(0)\, \delta^2 .$$

By (19) the central mass density is

$$\rho(0) = m\, n(0) = \frac{\mathfrak{M}}{2\pi \int S d\delta} \operatorname*{Lt}_{\delta \to 0} \frac{S(0) - S(\delta)}{\delta^2} . \tag{28}$$

The strip count data give

$$\rho(0) \cong 1.6 \times 10^{-25}\, h^2\ g\ cm^{-3} , \tag{29}$$

in reasonable agreement with the isothermal gas sphere model. The surface density (mean mass in unit column) may be estimated from Wilson's galaxy counts in annular rings.[12] After correction for the estimated density of background galaxies one finds that 4.3 percent of the galaxies are found in the central disc within a radius of 4' arc. On the assumption that the mass is distributed like the galaxies this is the fraction of the total mass (25) within this area. This gives

$$\sigma(0) \cong 0.6\ h\ g\ cm^{-2} . \tag{30}$$

According to this result, if the cluster were flattened down to the density of solid hydrogen ($\rho = 0.07$ g cm^{-3}) it would be about 10 cm thick at the middle.

The density (29) represents an extreme central concentration – by comparison if the mass (25) were uniformly distributed over a sphere of 100' arc radius, the nominal cluster radius, the resulting density would be 2 orders of magnitude smaller than (29). The estimated mean mass density

(6) is another 4 orders of magnitude smaller. It might be thought that (29) could be reduced by postulating more smoothly distributed "missing mass." The difficulty with this would be that the observed strong central concentration of galaxies could be preserved only if the velocity dispersion were smaller toward the center. The available data reveal no such effect, and suggest, if anything, a trend in the other direction. Put another way, the observed velocity dispersion across the face of the cluster is consistent with the assumption that the gravitational potential field has the shape implied if the mass density is proportional to the observed galaxy density in the cluster. This is illustrated in Figure IV-1.

Now let us consider some possibilities for missing mass. First, the value for the cluster luminosity $\mathcal{L}$ is based on a sum over the measured luminosities of the observed galaxies in the cluster. As in the discussion of L, in Section (b), we must recognize that this could be seriously in error because it misses the numerous very faint dwarf galaxies, or, for that matter, dispersed intracluster stars. We can check this by considering the expected surface brightness of the cluster. For example, let us suppose that the mass-to-light ratio of the cluster is in fact 20 h (eq. 5). Then the expected central surface brightness is

$$i(0) = \int \frac{L}{4\pi} dr$$

$$= \frac{(\mathcal{M}/\mathcal{L})^{-1}}{4\pi} \int \rho \, dr$$

$$= \frac{\sigma(0)}{4\pi} (\mathcal{M}/\mathcal{L})^{-1}.$$

With equation (30) this reduces to

$$i(0) \cong 600 \, S_{10} \, (V) \, ,$$

(eq. A-5) which is fully as bright as the night sky. Such a great brightness appears to be excluded even by visual inspection of photographic plates.

A more serious possibility is that an observational error in $\mathcal{L}$ might account for some fraction of the anomaly, bringing $\mathcal{M}/\mathcal{L}$ down to 150 h, say, which is on the upper end of what we might expect to be the mass-to-light ratio for elliptical and SO galaxies. There are several experiments in progress to answer the very important question whether the integrated cluster luminosity could be significantly larger than the present accepted value.

The next interesting possibility is intracluster gas. We have a very strong upper limit on the possible amount of atomic hydrogen from the lack of detectable 21 cm emission.[14] This result is not altogether surprising, for if hydrogen were present in any appreciable amount it should collisionally ionize itself in short order, making a plasma.

There are very strong objections to the assumption that the "missing mass" in the Coma cluster may be a plasma because of the expected rate of bremsstrahlung emission (radiation by electrons scattered in the field of the ions).[15] The first effect to worry about is the net energy dissipation. The characteristic cooling time for the plasma by bremsstrahlung emission is (eq. (67) below)

$$t_c = 9 \times 10^3\, T^{1/2}\, n_e^{-1}\ \mathrm{y},$$

where T is the plasma temperature (K) and n_e is the electron density = proton density (particles cm^{-3}). If the plasma is pure hydrogen, free-free emission is the dominant energy loss at temperature $\gtrsim 10^5$ K. If helium is present there is also a strong contribution due to electron collisional excitation of He I and He II unless $T \gtrsim 10^6$ K.

If the central plasma density is given by (29) the central particle density is

$$n_e \sim 0.1\, h^2\, \mathrm{cm}^{-3}.$$

The expected plasma temperature is fixed by the mean square velocity (17) if it may be assumed that the plasma scale height everywhere matches the scale height of the galaxy density,

$$T = m_H \langle v_{\parallel}{}^2 \rangle / (2\,k) \sim 6 \times 10^7\ \mathrm{K}. \qquad (31)$$

This gives

$$t_c \cong 8 \times 10^8 \ h^{-2} \ y,$$

one tenth of the expected cluster age, $\sim H^{-1} = 1 \times 10^{10} \ h^{-1} \ y$, an unhappy result because if the plasma temperature falls the gas kinetic pressure cannot support the plasma, the system contracts, the gas density rises, increasing the rate of loss of energy. The result evidently is a catastrophic collapse of the plasma. The situation might be saved by introducing an energy source, but there are problems with stability,[15] or by assuming a lower central density, but it was remarked that this goes against the self-consistent shape of the cluster.

A more direct test for an intracluster plasma is to look for the X-ray bremsstrahlung radiation by the plasma.[15] The greatest contribution would be from the more dense central part of the cluster. Judging from the distribution of galaxies the density does not appreciably fall from the central value out to a distance $s \sim 10'$ from the center, compared to the nominal cluster radius $R = 100'$ arc. On taking the plasma density within the sphere of radius s to be given by (29), and using the bremsstrahlung radiation formula (66) discussed below, one finds that the expected total X-ray flux per logarithmic bandwidth at energy E from the Coma cluster would be

$$\nu f_\nu \cong 17 \ h^3 \ T_8^{-1/2} \ e^{-E_{keV}/(8.6 \ T_8)} \ \text{photons cm}^{-2} \ \text{sec}^{-1},$$

where T_8 is the plasma temperature measured in units of 10^8 K.

If the plasma temperature were given by equation (31) the incident flux at 2 keV would amount to 10 photons cm^{-2} sec^{-1} per logarithmic energy interval, comparable to the powerful X-ray source Sco X – 1![16] It appears that the X-ray flux is well below three orders of magnitude less than this model prediction.[16] The first option then would be to assume that the plasma is much cooler than (31), perhaps distributed in wisps. The most direct difficulty with this is that the cooling time gets extremely short. One could then add heat sources, but it is very hard to see how the sources could be adjusted to keep the plasma at the right temperature, rather than blowing it out of the cluster, for example. A second option would be to reduce the central concentration of the plasma distribution. As always the objection to this is that where the mass appears to be missing is near the center of the cluster. A third option would be to accept that a plasma could not make any appreciable contribution to the cluster mass.

Even with the third option it would be of considerable interest to detect and trace out the shape and spectrum of X-ray emission from compact clusters like Coma. The galaxies presumably are shedding some gas, and this gas should be hot, $T \sim 10^8$ K (eq. 31) until it loses energy and settles to the potential minimum, which presumably is back in the central galaxies. It now appears that the Coma cluster is an X-ray source of about the right angular extent and spectrum to fit this picture.[16] If the source is a hot intracluster plasma, and if the plasma is smoothly distributed, as seems reasonable, then the plasma mass is estimated to be one percent of the mass (25).

A second conceivably observable effect of the plasma would be Thomson scattering of the light from background galaxies. If the surface density of the plasma were given by (30) the optical depth against Thomson scattering looking through the center of the cluster would be

$$\tau = \sigma_T \, \sigma(0)/m_H \sim 0.3\, h ,$$

where σ_T is the Thomson cross-section. This appears to be too small to be detectable.

A fascinating suggestion is that the missing mass in the Coma cluster may be found in condensed hydrogen "snow."[17] We can place an approximate lower bound on the possible size of such snow particles by observing that if the material were too finely divided it would excessively scatter light, obscuring visibility of distant galaxies behind the cluster. This very important consideration apparently was first mentioned by Zwicky in his discussion of intergalactic dust. Let us assume that each particle is spherical, radius r, with density ρ_I not greater than 0.07 g cm^{-3}, the density of solid hydrogen, and probably a good deal less because the snow might be expected to have a fairy-castle structure. If $10 \text{ cm} \gtrsim r >> \lambda$ the scattering cross-section of a grain is $\sigma_S = 2\pi r^2$, twice geometrical because diffraction scattering removes light from the beam. The optical depth looking through the cluster center is

$$\tau = \int \sigma_s n_s \, dx$$

$$= 2\pi r^2 \int \rho \, dx / \left(\frac{4}{3} \pi r^3 \rho_I\right)$$

$$= 3\,\sigma(0)/(2r\,\rho_I)\,,$$

where $\sigma(0)$ is given by (30). We can see distant galaxies behind the cluster, so τ could not greatly exceed unity, whence

$$r \gtrsim 10 \text{ h cm.}$$

That is, if we wanted to have a significant amount of mass in this form we would have to invoke rather substantial snowballs, not snowflakes, and it is difficult to imagine how they could have formed. One might have thought to avoid this by going to grains very much smaller than one micron, so the scattering cross-section is reduced from the geometrical by the factor $\sim (r/\lambda)^4$. The problem with this is that the cross-section for collisions of grains remains the geometrical cross-section. The velocity dispersion of the grains must be on the order of 1000 km sec^{-1}, for the grains otherwise would collect at the center of the cluster. It follows that the mean free time for collisions among the grains would be very short, much less than the crossing time if the grain radius $<< 10$ cm. Collisions at $\sim 1000 \text{ km sec}^{-1}$ being disastrous to the grains we conclude that the snow would be vaporized in short order.

A subject of great current interest is the gravitational collapse to black holes predicted in general relativity theory. Following Ruffini and Wheeler[18] one can say that these collapsed objects are black because no radiation can leave them and holes because matter and radiation can fall into them but cannot get out again. Black holes have gravitational mass, the mass m of a black hole being defined such that a particle in distant orbit around the object would move in the Newtonian orbit appropriate to an ordinary Newtonian gravitational mass m.

Small black holes dispersed among the galaxies could bind the cluster. Van den Bergh has pointed out that large black holes would not be acceptable because they would tidally disrupt the galaxies.[19] For example, let us suppose there is at the center of the cluster one massive black hole, mass $\mathfrak{M}_B = 1.5 \times 10^{15}\ h^{-1}\ \mathfrak{M}_\odot$ (eq. 25). The Roche limit for stability against tidal disruption of a galaxy with mass m and radius r a distance R away is

$$R > r\left(\frac{2\mathfrak{M}_B}{m}\right)^{1/3} . \tag{32}$$

Rood and Baum[20] give a useful list of apparent magnitudes and angular sizes to surface brightness ~ 25 photovisual magnitudes per square second of arc for galaxies in the central parts of the Coma cluster, from which we can get for each galaxy the minimum distance R from the black hole for stability against tidal disruption. The results are in the range

$$R > \alpha\ (h/f)^{1/3} \text{ min arc},$$
$$80 > \alpha > 20,$$

where f is the galaxy mass-to-light ratio in Solar units,

$$f = \left(\frac{\mathfrak{M}}{\mathfrak{M}_\odot}\right) / \left(\frac{\mathfrak{L}}{\mathfrak{L}_\odot}\right).$$

The largest value of α was for NGC 4889, the brightest centrally placed galaxy, the smallest α for one of the faintest in the list. If we adopt a conventional mass-to-light ratio, $f = 50h$, and take $\alpha \sim 40$, say, we find

$$R \gtrsim 10 \text{ arc min}.$$

By comparison Omer, Page and Wilson find that the number density of galaxies in the central parts of the cluster is[12]

$$n(0) \sim (5 \text{ arc min})^{-3}.$$

On this picture there ought to be conspicuous and ongoing disruption of galaxies, and as there is not we must conclude that the cluster could not be bound by one massive black hole.

Next, let us consider the idea that the cluster may be bound by numerous smaller black holes, each of mass $\mathfrak{M}_B$, distributed in space in the same fashion as the galaxies. When the black holes pass through galaxies they knock out stars. Because the typical collision velocity (17) is so much larger than the internal stellar velocities the simple impulse approximation implies, and it is readily seen that the fractional mass loss due to a collision between a hole and a galaxy is

$$\frac{\delta m}{m} \sim \frac{G\mathfrak{M}_B{}^2}{m\, v_B{}^2\, r},$$

where m and r are the mass and radius of the galaxy, as before, and v_B is the collision velocity $\sim 1000\ \text{km sec}^{-1}$. Suppose the mass (25) is supplied by $\mathfrak{M}/\mathfrak{M}_B$ black holes. To get at the collision rate let P(R) be the spatial distribution function for galaxies and for black holes, normalized to $4\pi \int R^2\, P(R)\, dR = 1$. Then in time H^{-1} the mean fractional loss of mass from a galaxy due to collisions with black holes is

$$\frac{\delta m}{m} \sim \int 4\pi\, R^2\, P(R)\, dR \left(\frac{G\mathfrak{M}_B{}^2}{m\, v_B{}^2\, r}\right)(\pi\, r^2)\left(\frac{\mathfrak{M}}{\mathfrak{M}_B} P(R)\, v_B\right) H^{-1}\ .$$

The first term in parentheses is the fractional mass loss per collision, the second is the collision cross-section, and the third is the flux of black holes. The integral averages the loss rate over all positions for the galaxy. This fraction should be less than about unity if the cluster age is comparable to H^{-1}. The implied limit on $\mathfrak{M}_B$ is

$$\mathfrak{M}_B \lesssim 10^9\, \mathfrak{M}_\odot f\, h^{-2}\ ,$$

where again f is the mass-to-light ratio for the galaxies and m and r are taken from the smaller of the galaxies in the central parts of the cluster. It appears from this limit that the black holes could be as massive as the more modest galaxies without doing appreciable damage.

What are we to conclude about the mass of the Coma cluster? First, we cannot be sure that there is an anomaly because the statistics are limited. The indicated mass-to-light ratio is $\sim 400\, h$, the conventional mass-to-light ratio f is perhaps $50\, h$, the anomaly a factor of 8. However, the statistical uncertainty in $\mathfrak{M}$ (25) is a factor of 2,[6] and probably there is a like uncertainty both in $\mathfrak{L}$ and in the "expected" f value.

It will be of considerable interest not only to accumulate better statistics on the Coma cluster but, as we recognize that the statistics on this one cluster are inherently limited, also to have comparable data on other rich compact clusters.

If there is an anomaly it is by no means clear how we will interpret it. We have argued that it is difficult to see how there could be a dynamically interesting amount of dispersed intracluster matter, and of course it will be of continuing interest to extend and improve these tests. Black holes of sufficiently small size could provide the missing mass. The distressing prospect here is that if they are small enough we would seem to have little or no chance to test the hypothesis in an independent way.

ii) *The M81 Group*

The M81 group of galaxies is roughly similar to the Local Group in mass and extent, and aside from the Local Group it is the best studied of the small groups of galaxies. It is also a good illustration of the ambiguities one encounters in trying to understand the dynamics of such systems.

Table IV-1

M81 GROUP

Member	Recession Velocity (km sec^{-1})	Corrected Velocity	θ (min. arc)	Min. Mass $10^{10}\mathfrak{M}_\odot$
M81	–40	76	–––	–––
M82	280	398	38	43
NGC 3077	–41	74	46	–––
NGC 2976	42	153	83	5
IC 2574	46	162	180	15
NGC 2403	132	239	810	240
NGC 2366	110	231	770	200

This group is 3.25 Mpc away according to the distance found by Tammann and Sandage for an outlying member (NGC 2403; ref. III-5).

Radial velocities of the main group members are listed in Table 1. The second column is the observed apparent recession velocity, as usual the negative numbers meaning blueshift. All the velocities are taken from a survey by Roberts[21] with the exception of NGC 3077, taken from the recent study by Demoulin,[22] and NGC 2976, taken from the Humason, Mayall and Sandage compendium (ref. II-16). In the third column the observed velocity is corrected for the component of our rotation velocity looking in the direction of the galaxy, the assumed rotation velocity being 250 km sec^{-1}. The fourth column is the angular distance of each galaxy from M81, which is the brightest object in the system.

The first five galaxies in the list form a sub-group with projected diameter 220 kpc if the distance is 3.25 Mpc. The last two galaxies are fairly close together and some distance from the sub-group.

Because M81 is by far the brightest of the galaxies one might assume that it provides most of the mass. Then for each of the other galaxies one can find a minimum value of this mass on the assumption that the galaxy is gravitationally bound to M81,

$$\mathfrak{M} \geq r\, v^2/2G\ ,$$

where r is the projected distance and v the line of sight velocity relative to M81. Projection effects can only reduce both of these factors. The minimum masses derived in this way are listed in the fifth column of the table. By comparison the luminosity of M81 is $1 \times 10^{10}\, \mathfrak{L}_\odot$, and assuming a mass-to-light ratio of 10 this would make

$$\mathfrak{M} \sim 1 \times 10^{11}\, \mathfrak{M}_\odot .$$

Three of the members of the inner sub-group agree with this mass, the fourth, M82, conflicts with it. We know that M82 is a member of the group, not a chance background object, because Roberts (ref. I-46) has been able to trace a bridge of 21 cm emission from M81 to M82 (and to NGC 3077). What is more the radial velocity in the bridge shows a systematic gradient going from M81 to M82, running more or less smoothly between the estimated radial velocities of the two galaxies. However, it is still quite possible that the radial velocity of M82 listed in the Table has an appreciable error because this is a complicated object, that apparently has suffered a violent explosion, so it is hard to know how to relate the observed motion to the motion of the center of mass. It will be of considerable interest to see whether the detailed motions in the HI bridge can be used to substantiate the high relative velocity of M81 and M82, or whether this information can be used to argue that the true relative velocity may be smaller than the value assumed here.

The two galaxies NGC 2403 and 2366 are close together in the sky, have similar line of sight velocities, and are well removed from the rest of the group. The enormous mass needed to bind them to the rest of the group is based on the assumption that the velocity of the center of mass of the inner sub-group is roughly equal to the motion of M81, which is reasonable if the mass-to-light ratio is constant from galaxy to galaxy. The more attractive ways to get around this are (1) NGC 2403 and 2366 are an independent pair appearing as a chance projection near the M81 group; (2) the estimated motion of the center of mass of the inner sub-group is in error because M81 contains half the total mass, M82 most of the remainder; (3) there is a very large amount of "missing mass" in the group; (4) the group is localized in space but dissolving.

iii) *The Local Group*

Very pleasant descriptions of the Local Group have been given by Hubble (ref. I-1) and more recently by van den Bergh (ref. III-2). The problem here is not with the balancing of kinetic and gravitational energies but with the time scale. The argument was first stated by Kahn and Woltjer,[23] and in modified form goes as follows.

Until recently the Local Group was thought to consist of two well-defined subgroups, one around the Galaxy and the other around M31 (the Andromeda Nebula), plus one or two outliers. The Galaxy and M31 being the dominant systems we can therefore treat the Local Group as a 2-body problem. The present separation is put at

$$\ell_0 = 690 \text{ kpc}, \tag{33}$$

and the relative velocity is

$$\frac{d\ell_0}{dt} = -290 + 250 \times 0.80 = -90 \text{ km sec}^{-1} . \tag{34}$$

The first term is the mean of the observed optical and 21 cm velocities, in the second term our assumed rotation velocity in the Galaxy is 250 km sec^{-1}, and the projection factor in the direction of M31 is 0.80. According to the standard Big Bang cosmology it would be supposed that in the early Universe the Galaxy and M31 were close together and moving radially apart according to the law of the general recession, $v = (\dot{a}/a)\,\ell$, where the constant of the day, $\dot{a}/a$, would be larger than it is now. One would suppose that M31 is moving toward us now because the gravitational attraction between the two bodies eventually slowed and reversed the initial expansion. We can find a minimum mass of the system by assuming that we are just coming together on the initial collapse, and by requiring that this has happened in the available time, $\sim 10^{10}$ y. Then the separation between the two masses satisfies the ordinary 2-body equation

$$\frac{d^2\ell}{dt^2} = -\frac{G\mathfrak{M}}{\ell^2}, \tag{35}$$

where $\mathfrak{M}$ is the sum of the masses of the two galaxies. The boundary conditions are (1) at $t \to 0$, $\ell \to 0$, the galaxies starting out close together; (2) at the present time, $t = 10^{10}$ y, say, ℓ and $d\ell/dt$ satisfy equations (33) and (34). These conditions overconstrain the differential equation, and this fixes the system mass $\mathfrak{M}$. The solution to the differential equation (35) is the parametric form $\ell = A(1 - \cos\eta)$, $t = B(\eta - \sin\eta)$. The boundary values give the present value of η as 4.0 radians, the maximum separation was $2A = 840$ kpc, and this happened at time $\pi B = 6.6 \times 10^9$ y (assuming the present time is 1×10^{10} y). The mass is

$$\mathfrak{M} = 3.7 \times 10^{12}\,\mathfrak{M}_\odot . \tag{36}$$

The most uncertain parameter in the model is the time. If t were doubled the mass would be reduced by a factor 1.7.

The luminosity of M31 is $2.1 \times 10^{10}\ \mathfrak{M}_{\odot}$, and the contribution from the Galaxy is somewhat smaller, so the indicated mass-to-light ratio for the Local Group is about 100 Solar units. This is a factor of 100 larger than the mass-to-light ratio of material seen in the Solar neighborhood, and it is a factor of about 10 larger than the typical value for giant spiral galaxies derived from rotation curves.

As usual there are a number of possible interpretations of this discrepancy: (1) It has been pointed out that the M31 sub-group and the Galaxy sub-group may simply be independent systems that happen to be passing by. This requires the coincidence that the groups should be close together as we now happen to observe them. (2) Kahn and Woltjer suggested that the missing mass may be supplied by matter between the galaxies in the Local Group. Oort has revived this thought, and suggests that the intra-group matter may be the source of the high velocity HI clouds.[24] (3) We do not understand very well the masses of galaxies. If galaxies have approximately spherically symmetric halos the rotation curve at any point does not say anything about the mass in the halo outside that radius. The only direct test is to ask whether the rotation velocity in the outer parts of the disc varies with radius as $r^{-1/2}$, as would be expected if all the mass were concentrated well within r. This is hard to establish to any precision, and of course it is still only an extrapolation. (4) There may be another massive member of the Local Group, Maffei 1.[25] It is too soon to say whether this object could solve the problem because it is so highly obscured by the Galaxy that it is hard to be sure what it is. The present indication is that the mass of Maffei 1 could be as large as the missing mass (36) only if it were some 4 Mpc away, which would seem to be too far to do much good.

iv) *Seyfert's Sextet*

Yet another variant of the stability problem is found in compact groups of galaxies, of which Seyfert's Sextet is an example. As the name would suggest this group was discovered by Seyfert.[26] His paper contains a 200'' direct photograph of the system taken by Baade. As the name also would suggest there are six objects in the group, three of which seem to be spiral galaxies, one a low surface brightness cloud of some sort. The redshifts of the five brighter objects were found by Sargent, and the results discussed by Sargent and Burbidge.[27] Four of the five redshifts are close to 4400 km sec^{-1}, but the fifth, for one of the spiral galaxies, is 19930 km sec^{-1}.

The conventional interpretation is that this anomaly would have to be the result of a chance coincidence with a distant galaxy of high redshift, and in fact the *a priori* probability that this might have happened is not manifestly unreasonable. The maverick has apparent photographic magnitude ~ 16. According to Hubble there are $n \sim 3$ galaxies per square degree this bright or brighter. Seyfert's sextet has a diameter of 1 arc minute, or area $A \sim 2.5 \times 10^{-4}$ square degrees. The *a priori* probability of a coincidence with a background galaxy brighter than $m \sim 16$ is therefore $nA \sim 8 \times 10^{-4}$. This is the probability of a coincidence if an object were chosen *at random* and examined within this area A. However Seyfert surely did not choose his object at random – he picked it out because it looked peculiar. The appropriate statistic therefore is the ratio of the number of objects in the sky that look peculiar because of chance coincidence to the number of "peculiar objects" found in a complete sky survey. As we apparently do not have a systematic sky survey of this sort the best that can be done is to ask for the number of chance coincidences there are in some fraction f of the total sky among randomly distributed galaxies brighter than apparent photographic magnitude 16. If the search area is F square degrees the expected number of single coincidences within the area A of Seyfert's Sextet is $n^2AF/2 \sim 50f$, where f is the fraction of the whole sky surveyed. The number of double coincidences, three unrelated objects within area A, is $n^3 A^2F/6 \sim 0.014f$. Apparently we can make a strong *a priori* statement against the probability of finding a double coincidence even if the whole sky were searched; but we cannot make a similar *a priori* statement against a single coincidence.

If the maverick is ignored Seyfert's system still is very peculiar. The mean redshift implies that the distance is $44\ h^{-1}$ Mpc, which makes the diameter of the group ~ 14 kpc, only about half the diameter of the Galaxy.[27] The luminosity of the brightest member is $3 \times 10^9\ h^{-2}\ \mathcal{L}_\odot$, about one third of the luminosity of the Galaxy, but its diameter is only $4.9\ h^{-1}$ kpc. Application of the virial theorem yields group mass-to-light ratio ~ 75 h,[27] which is high but perhaps may not be considered a serious anomaly because the brighter two galaxies are peculiar systems so they need not necessarily have mass-to-light ratios like more familiar objects. The crossing time, defined as the projected diameter divided by the dispersion of radial velocities, is $\sim 3 \times 10^7\ h^{-1}$ y, which means ~ 300 crossing times in a Hubble time H^{-1}. It is difficult to see how the two spiral galaxies in the group could have avoided colliding for so long, or, if they did collide, how they could have avoided losing their spiral structure. This is a reverse of the Baade-Spitzer argument,[28] that compact clusters contain no spiral galaxies because collisions among galaxies have stripped the spirals of their gas and dust, leaving SO systems. Here there seems to be ample time for the effect but it apparently did not happen. An interesting point made by Baade (quoted by Seyfert)[26] is that one of the spirals, Seyfert's object "c," seems to be in danger of tidal disruption. On using the projected distance to the brightest member one finds that for stability where it is now "c" must have mass-to-light ratio ~ 20 h, which is suspiciously large for a spiral galaxy, and of course it must be assumed that "c" never got much closer to the dominant members.

Other examples of compact clusters are Stefan's quintet and VV 172, in both of which one member has redshift wildly different from the rest.[29] It is by no means clear what these groups are telling us, but there are some evident questions: (1) Are compact groups, peculiar in one way or another, more common than would be expected from projection effects (chance coincidences plus elongated systems viewed from the appropriate direction)? (2) Are compact groups stable for $\sim 10^{10}$ y against the ordinary gravitational relaxation processes that tend to evaporate some members and make others come together?[30] (3) Are the members of a small group stable against tidal disruption or coalescence? For this last question it will be noted that the situation is quite different from the great clusters, for here the relative velocities are just comparable to internal velocities, which ought to favor disruption and tidal drag.

v) *Summary*

These examples illustrate our profound lack of understanding of the elements of galaxy dynamics, and there are many other examples.[31] It seems clear that there is a lot to learn from close study of the stability problem. It is less clear whether the lesson will be that we have got the laws of physics wrong, or only that the complexity of the astronomical Universe has left enormous room for ambiguous observations and misleading interpretations.

d) *Intergalactic Matter*

The spaces between the galaxies provide a good deal of room for gas, and a good deal of effort, theoretical and experimental, has been devoted to the possible manifestations of any such material. Following the usual assumption in cosmology, it is supposed that the matter is distributed in a smooth homogeneous fashion. It is commonly assumed also that the intergalactic material has not suffered cycling through stars. Since the oldest known stars in our Galaxy appear to be nearly pure hydrogen (with perhaps helium, which adds little to the problem), it would follow that the intergalactic material is hydrogen in some form (atomic, molecular, ionized, condensed "snow").

The first assumption is one mainly of convenience. If taken literally it is somewhat forced because the galaxies are known to be distributed in a decidedly irregular fashion up to dimensions of at least 10 Mpc. How did the intergalactic hydrogen escape the same fate? Two answers are possible. (1) It just did. (2) The hydrogen did get concentrated into lumps like groups and clusters of galaxies, but the lumps failed to fragment into visible stars ("unborn galaxies"). With the second answer the usual analysis may have to be modified because the material is optically thick, and there are problems with stability over 10^{10} y.[32]

It would not be surprising to find that intergalactic gas is distributed in wispy or cloud-like structure, like the galaxies, but the following discussion will be almost entirely under the assumption that the density distribution is close to homogeneous. Analyses of more complicated situations clearly are needed but tedious because they degenerate into questions of detailed discussions of special cases.

i) *HI-21 cm Emission*

This test is based on the expected 21 cm emission due to transitions between hyperfine levels in the ground state of atomic hydrogen. It was first applied by Goldstein.[33]

The hyperfine splitting in atomic hydrogen arises from the coupling of the proton and electron magnetic moments, the higher energy state having spin angular momentum $F = 1$, the lower $F = 0$. The spontaneous decay rate and the wavelength for the transition are

$$\Lambda_1 = 2.85 \times 10^{-15} \text{ sec}^{-1}, \qquad \lambda_1 = 21.1 \text{ cm}. \tag{37}$$

To get at the expected radio background from intergalactic hydrogen we need to know the rate of emission of radiation per unit volume, which in turn depends on the relative numbers of atoms in ground and excited states, n_1 and n_2. It is conventional to specify this in terms of an equivalent spin temperature T_s by the usual Boltzmann formula

$$\frac{n_2}{n_1} = 3e^{-h\nu_1/kT_s}, \tag{38}$$

the factor of 3 arising because there are 3 excited states, one ground state. When $T_s >> h\nu_1/k = 0.07$ K (38) says that 3/4 of the atoms are in the excited state, so if there are n atoms per unit volume the rate of spontaneous radiative transitions to the ground state is $(3/4)\,n\Lambda_1$ per unit volume, and the volume luminosity is

$$j_\nu = \frac{3}{4}\frac{n\Lambda_1 h \nu_1}{4\pi}\,\delta(\nu - \nu_1) \text{ ergs cm}^{-3}\text{ sec}^{-1}\text{ Hz}^{-1}\text{ ster}^{-1}, \tag{39}$$

where the shape of the line has been approximated as a Dirac delta function. It then follows from equation (12) that the contribution to the brightness of the isotropic radio background by 21 cm emission is

$$\begin{aligned} i(\nu, t_0) &= \int_0^{t_0} \left[\frac{a(t)}{a_0}\right]^3 j(\nu\, a(t_0)/a(t), t)\, c\, dt \\ &= \frac{3}{4}\frac{n(HI)\,\Lambda_1 h\, \nu_1 c}{4\pi} \int_0^{t_0} \delta(\nu\, a_0/a - \nu_1)\, dt. \end{aligned} \tag{40}$$

In the second equation it is assumed that $n(t) = n(HI)\,(a_0/a(t))^3$, $n(HI)$ being the present density of atomic hydrogen. The integral vanishes when $\nu > \nu_1$ and is equal to $[\nu_1\, \dot{a}(t)/a(t)]^{-1}$ when $\nu < \nu_1$. That is, there is a step in the radio brightness at 21 cm wavelength, the step in brightness amounting to

$$\delta i(\nu_1) = \frac{3}{4}\frac{n(HI)\,\Lambda_1 h\, c}{4\pi H}.$$

Radio brightness conventionally is expressed in terms of the equivalent antenna temperature T_a, defined by the linear formula

$$i(\nu) = 2k\, T_a(\nu)\, \nu^2/c^2.$$

This form is the familiar Rayleigh-Jeans approximation to the long wavelength end of a blackbody spectrum. The discontinuity in brightness temperature is then

$$\begin{aligned} \delta T_a &= \frac{c^2}{2k\,\nu_1^{\,2}} \cdot \frac{3n(HI)\,\Lambda_1 h\, c}{16\pi\, H} \\ &= 1.68 \times 10^4\, n(HI)\, h^{-1}\text{ cm}^3. \end{aligned} \tag{41}$$

The best upper limit on the possible size of δT_a is due to Penzias and Wilson,[34] who find

$$\delta T_a < 0.08 \text{ K}.$$

By equation (41),

$$\begin{aligned} n(\mathrm{HI}) &< 5 \times 10^{-6}\ h\ \mathrm{cm}^{-3}, \\ \rho(\mathrm{HI}) &< 8 \times 10^{-30}\ h\ \mathrm{gm\ cm}^{-3}, \\ \rho(\mathrm{HI})/\rho_c &< 0.4\ h^{-1}. \end{aligned} \tag{42}$$

It has been assumed here that $T_s >> 0.7$ K and that self-absorption by the atomic hydrogen may be neglected. The first assumption is reasonable because the spin temperature is not expected to fall below the ambient radiation temperature, 2.7 K (Chap. V). The second assumption is violated unless $T_s >> 2.7$ K, for if $T_s = 2.7$ K the spin system is in equilibrium with the radiation, hence there cannot be a net energy transfer, while if $T_s < 2.7$ K the matter must absorb energy from the hotter radiation, producing a *decrease* in the background going longward of 21 cm wavelength.

ii) *21 cm Absorption*

The second test for intergalactic atomic hydrogen is to look for absorption in the radio spectrum of an extragalactic radio source. This test was first applied by Field.[35]

One can compute the opacity of the atomic hydrogen knowing the emission j_ν by a simple thermodynamic argument. The brightness of radiation moving along the path $x = ct$ varies as

$$\frac{\partial i_\nu}{\partial t} = (j_\nu - \kappa_\nu\, i_\nu)\, c, \tag{43}$$

where the first term represents augmentation of the brightness due to emission, the second attenuation due to absorption, the coefficient of opacity being κ. The functions j and κ depend on the relative population of ground and excited states, n_2/n_1, which in turn is fixed by the spin temperature T_s. Now clearly if the brightness i_ν in equation (43) near 21 cm wavelength were equal to the brightness of blackbody radiation at the spin temperature T_s the hydrogen in effect would be in thermal equilibrium with the radiation field, and for this choice of i_ν emission and absorption rates must just balance. That is,

$$j_\nu = \kappa_\nu P_\nu , \tag{44}$$

where P_ν is the Planck blackbody radiation brightness at temperature T_s,

$$P_\nu = \frac{2h\,\nu^3}{c^2(e^{h\nu/k\,T_s} - 1)} .$$

In equation (44) we need an expression for $j(\nu)$ more general than (39), where it was assumed $n_2/n_1 = 3$, that is, T_s very large. More generally by (38) we have

$$n_2 = 3n\, e^{-h\nu_1/k\,T_s} / (1+3e^{-h\nu_1/k\,T_s}),$$

$$j_\nu = \frac{3\Lambda_1 n\, h\nu_1}{4\pi}\,\delta(\nu-\nu_1)\,\frac{e^{-h\nu_1/k\,T_s}}{1+3e^{-h\nu_1/k\,T_s}},$$

$$\kappa_\nu = j_\nu/P_\nu$$

$$= \frac{3}{8\pi}\left(\frac{n}{1+3e^{-h\nu_1/k\,T_s}}\right)\Lambda_1\left(\frac{c}{\nu_1}\right)^2 \delta(\nu-\nu_1)\,(1-e^{-h\nu_1/k\,T_s}) . \tag{45}$$

The last term in parentheses represents the correction factor for stimulated emission. When $h\nu_1 << k\,T_s$, (45) becomes

$$\kappa_\nu = \frac{3}{32\pi}\,\frac{n\Lambda_1 h\,\nu_1}{k\,T_s}\left(\frac{c}{\nu_1}\right)^2 \delta(\nu-\nu_1) . \tag{46}$$

The brightness of the radiation coming from a distant radio source satisfies the equation (cf. eq. 11)

$$\frac{d}{dt} i(t, \nu(t)) = -3i\frac{\dot a}{a} - \kappa(t, \nu(t))\, c\, i . \tag{47}$$

We are neglecting here the emission term, which generates the isotropic background. The solution to equation (47) is

$$i(t_0, \nu) = i(t_i, \nu_i)\,(a_i/a_0)^3\, e^{-\tau(\nu)},$$

where the subscript i refers to initial values at the source, and τ is the optical depth,

$$\tau(\nu) = \int_{t_i}^{t_0} \kappa(t, \nu(t))\, c\, dt . \tag{48}$$

The optical depth vanishes unless the radiation frequency $\nu(t)$ passes through the resonance ν_1 somewhere between t_i and t_0,

$$\nu_1 > \nu > \nu_1 \, a(t)/a_0 ,$$

where ν is the frequency at which one is observing. By equations (46) and (48) the optical depth just longward of 21 cm wavelength is

$$\tau = \frac{3}{32\pi} \frac{\Lambda_1 h c^3 n(HI)}{k T_s \nu_1^2 H} = 1.68 \times 10^4 T_s^{-1} h^{-1} n(HI) , \qquad (49)$$

which agrees with (41).

Penzias and Scott[36] find $\tau < 5 \times 10^{-4}$ in the spectrum of the extragalactic source Cygnus A. By (49),

$$n(HI) < 3 \times 10^{-8} T_s \, h \, cm^{-3} . \qquad (50)$$

A similar limit is obtained by Allen.[36]

Now we must estimate the spin temperature T_s. Field[35] has shown that the spin temperature is determined by equilibrium among the rates of three processes: (1) radio frequency transitions between hyperfine states in the presence of the ambient radiation field at 21 cm wavelength; (2) collisional excitation and de-excitation by atoms and electrons; (3) resonance absorption of ultraviolet radiation.

The ambient radiation temperature at 21 cm wavelength is now known to be $T_0 = 2.7$ K, (Chapter V) and in the absence of any other perturbations the spin temperature would relax to this value in 10^7 y. Assuming a spin temperature of 2.7 K, (50) gives $n(HI) < 8 \times 10^{-8} \, h \, cm^{-3}$, well below the Einstein-deSitter density.

With the very low limit (50) on n/T_s now available one finds that if the hydrogen is uniformly distributed collisional excitation does not appreciably perturb the spin temperature. (For details see Field's second paper in ref. 35, especially his Figure 1. The only thing that has changed is the radiation temperature at 21 cm wavelength, which Field assumed to be 0.4 K, and we now know to be $T_0 = 2.7$ K, the effect being to move the asymptote at low n(HI) up by a factor 2.7/0.4.)

Lyman α resonance scattering can take an atom from the singlet ground state, say, to the 2P level, the atom decaying back down to the triplet state. The effect on the spin temperature depends on the spectrum of the radiation near 1215 A wavelength. To see what this spectrum must look like note that a photon near 1215 A changes frequency via three main processes, each with a very different time scale. The first effect is due to the motion of the atoms. If the matter temperature is T_K the typical speed of a hydrogen atom is $v \sim (3k\, T_K/m_H)^{1/2}$, and the frequency shift in a resonance scattering is $\delta\nu \sim \nu v/c \sim \nu(3k\, T_K/m_H\, c^2)^{1/2} \gtrsim 2 \times 10^9$ Hz if $T_K \gtrsim 2.7$ K. The scattering cross-section on resonance is $\sigma_0 = 7 \times 10^{-11}\ \text{cm}^2$. Random motion of the atoms smears out the resonance, reducing the effective cross-section at the peak to $\sim \sigma_0\, \Lambda_\alpha/(4\delta\, \nu) = 8 \times 10^{-12}\, T_K^{-1/2}\ \text{cm}^2$, where $\Lambda_\alpha = 6 \times 10^8\ \text{sec}^{-1}$ is the spontaneous decay rate. If the atomic hydrogen density were $10^{-5}\ \text{cm}^{-3}$ the mean free time for a photon with frequency within the Doppler-broadened resonance would be $\sim (8 \times 10^{-12}\, T_K^{-1/2} \times 10^{-5} \times 3 \times 10^{10})^{-1} \sim 4 \times 10^5\, T_K^{1/2}$ sec. The resonant scattering process can take an atom from the ground to the excited hyperfine state, or, with almost identical probability, can take an atom the other way. The difference in rates must balance the net rate of 21 cm radiative transitions. By (37) this rate can be as large as $\Lambda_1\, (kT_0/h\, \nu_1) = 1.1 \times 10^{-13}\ \text{sec}^{-1}$, where the second factor is the correction for stimulated emission (eq. 52 below). The reciprocal of this number is the second characteristic time, $\sim 10^{13}$ sec. A third characteristic number is $H^{-1}\, \delta\nu/\nu \sim 3 \times 10^{11}\, T_K^{1/2}$, the mean time for a photon to be redshifted through the resonance.

Of these processes the scattering time is by far the shortest so it has to be the dominant effect. In the scattering process the radiation acts like a classical relativistic Boltzmann gas, because stimulated emission is quite unimportant and photons are conserved to a good approximation. What is happening is that the part of the radiation gas with energy within the scattering resonance is very strongly coupled to the atomic hydrogen gas, in the sense that the photons within the resonance can be reshuffled in a time much shorter than any other process. We know from statistical mechanics what the result of this reshuffling must be – the part of the photon gas within the resonance and the hydrogen gas both relax to Maxwellian velocity distributions at a common temperature, T_K, say. The spin system in the hydrogen ground state is in thermal contact with the photon gas at effective temperature T_K (which of course is not a brightness temperature) and with the microwave radiation at $T_0 = 2.7$ K, and it finds its own equilibrium between the two.

To carry through this idea it is convenient to represent the brightness of a radiation field, $i(\nu)$ (units ergs cm^{-2} sec^{-1} $ster^{-1}$ Hz^{-1}) in terms of the effective number of photons per radiation mode and unit volume. Consider a large volume V, a frequency ν, a bandwidth $d\nu$, and a chosen direction of propagation within solid angle $d\Omega$. The number of radiation modes in this frequency range and moving into $d\Omega$ that can be fitted into V is (taking account of the two polarizations)

$$dN = 2V\,\nu^2\,d\nu\,d\Omega/c^3.$$

This number multiplied by the number $\mathfrak{N}$ of photons per mode times the energy per photon will be equal to the observed amount of radiation in the element $V\,d\Omega\,d\nu$,

$$(i/c)\,V\,d\Omega\,d\nu = (2Vh\,\nu^3/c^3)\,d\nu\,d\Omega\,\mathfrak{N},$$

or

$$\mathfrak{N}(\nu) = i(\nu)\,c^2/(2h\,\nu^3). \tag{51}$$

For blackbody radiation at temperature T_0,

$$\mathfrak{N}_{th} = (e^{h\nu/kT_0} - 1)^{-1}.$$

It will be recalled that Λ_1 is the rate of spontaneous decay of a hydrogen atom from the excited hyperfine state (eq. 37). In the presence of radiation the rate of decay has to be

$$\Lambda_1(1 + \mathfrak{N}_1)\,, \tag{52}$$

where the second term represents stimulated emission, $\mathfrak{N}_1$ being the number of photons per mode at 21 cm wavelength. The rate for absorption of photons by a hydrogen atom in the ground state is

$$3\Lambda\,\mathfrak{N}_1 \,. \tag{53}$$

The factor of 3 enters because there are 3 excited states belonging to spin $F = 1$, one ground state with $F = 0$. These results follow by the same detailed balance argument (of Einstein) used to derive equation (45).

We can also express the radiation spectrum near 1215 A in terms of the function $\mathfrak{N}(\nu)$. The photons near Lyman α resonance are being reshuffled among themselves at an enormous rate (relative to any other process of interest) so the photon distribution function must relax to statistical equilibrium with the matter. Because $\mathfrak{N} << 1$ the equilibrium is the Boltzmann distribution

$$\mathfrak{N} = \beta\, e^{-h\nu/k\,T_K} , \tag{54}$$

where β is a constant and T_K is the matter temperature. That is, there is imposed a small wiggle in the radiation spectrum at 1215 A.

Now we can compute the spin temperature T_S. The number of hydrogen atoms per unit volume in the ground state satisfies

$$\frac{dn_1}{dt} = \Lambda_1(1+\mathfrak{N}_1)\,n_2 - 3\,\Lambda_1\,\mathfrak{N}_1\,n_1$$

$$+ \frac{3}{4}\Lambda_\alpha\,\mathfrak{N}_\alpha\,n_2 - \frac{9}{4}\Lambda_\alpha\,\mathfrak{N}_\alpha\,e^{-h\nu_1/k\,T_K}n_1 \,. \tag{55}$$

The first term on the right hand side is the rate of spontaneous plus stimulated 21 cm transitions from the excited hyperfine state (eq. 52), and the second term is the rate of 21 cm excitation from the ground state (eq. 53). The third term represents the rate of transitions from the excited to the ground state via the Ly α resonance scattering process. The decay rate of an atom in the 2P state is

$$\Lambda_\alpha = 6.25 \times 10^8 \text{ sec}^{-1}, \tag{56}$$

and $\mathfrak{N}_\alpha$ is the mean number of photons per mode at wavelength $\lambda_\alpha = 1215$ A. The numerical coefficient may be understood as follows. The probability per unit time that an atom in the excited hyperfine state will absorb a Ly α photon is $3\Lambda_\alpha \mathfrak{N}_\alpha$, the factor of 3 because we are going from $L = 0$ to $L = 1$, and the mean probability that the atom will decay down to the $F = 0$ state is ¼. In the last term in (55), representing the reverse process, we have from (54) that the photon number is reduced by the factor $e^{-h\nu_1/k T_K}$ from the photon number in the third term, going the other way.

Because the relaxation rates in (55) all are much less than the evolution time H^{-1} we can set $dn_1/dt = 0$, solve for n_2/n_1, and set the result equal to the defining equation (38) for T_s. It can be assumed that T_0 and T_s both are much greater than $h\nu_1/k = 0.07$ K, the former because we now believe that the effective radiation temperature is 2.7 K, the latter because if it were not the limit (50) would be a very good one. Then one finds

$$T_s^{-1} = \left(\frac{1}{T_0} + \frac{\alpha}{T_K}\right) / (1 + \alpha),$$
$$\alpha = 3\Lambda_\alpha \mathfrak{N}_\alpha h\nu_1 / (4\Lambda_1 k T_0). \tag{57}$$

The best empirical limit on the extragalactic ultraviolet radiation background apparently is that of Kurt and Sunyaev.[37] Their limit on the flux in the bandwidth 1225 – 1340 A is

$$f \lesssim 2.5 \times 10^{-7} \text{ ergs cm}^{-2} \text{ sec}^{-1} \text{ ster}^{-1}.$$

Assuming an approximately flat spectrum this translates to

$$\nu\, i_\nu \lesssim 2.6 \times 10^{-6} \text{ ergs cm}^{-2} \text{ sec}^{-1} \text{ ster}^{-1}$$

just longward of Lyman α. More recently Lillie obtained the limits $\nu i_\nu < 4 \times 10^{-5}$ ergs cm^{-2} sec^{-1} ster^{-1} at $\lambda = 1400$ A, $\nu i_\nu < 2 \times 10^{-6}$ at 2400 A.[37] If we adopt the smaller limit, then we find from (51) and (57)

$$\alpha \lesssim 20.$$

It follows from (57) that for any T_K

$$T_s \leq (1 + \alpha)\, T_0 \lesssim 60 \text{ K}.$$

This result with (50) yields

$$n(\mathrm{HI}) \lesssim 1.7 \times 10^{-6}\ h\ \mathrm{cm}^{-3},$$
$$\rho(\mathrm{HI})/\rho_c \lesssim 0.15\ h^{-1}. \tag{58}$$

Apart from the ultraviolet flux limit, perhaps the most questionable assumption behind the limit (58) is that the hydrogen is reasonably uniformly distributed. If the hydrogen were in clouds (58) might be increased because collisional excitation could increase the spin temperature, and also because equation (49) is based on the assumption that the hydrogen is optically thin at 21 cm wavelength. If the hydrogen were in optically thick clouds it would cause a series of deep absorption lines, but the reduction in *mean* intensity would be $1 - e^{-\tau}$ which is less than τ for optically thick clouds.[32]

iii) *The Gunn-Peterson Test*

The idea here is to look for the attenuation of ultraviolet radiation from an extragalactic source due to scattering at the 1215 A Lyman α resonance in atomic hydrogen, and in the Lyman band in molecular hydrogen. The test is just the same as the search for attenuation of radio radiation due to scattering in the 21 cm transition. Just as before, for atomic hydrogen one looks for attenuation in the wavelength range (in Angstroms)

$$1215 < \lambda < 1215\,(1+z), \tag{59}$$

where z is the redshift of the source (eq. I-16). As will be seen this test is very much more sensitive than the analogous 21 cm test. It has been applied only very recently because the atmosphere is opaque to ultraviolet radiation shortward of about 3000 A wavelength. The breakthrough came not from space technology but from the discovery of quasars with redshifts $z > 2$, which brings the upper limit in (59) up past the atmosphere cut-off, so it can be observed from the ground.

As is so often the case this brilliant trick was "discovered" a number of times, by Shklovsky, Scheuer, and Gunn and Peterson, among others.[38,39] Gunn and Peterson had the great good fortune to have made the suggestion at the right time and place, at Caltech, when Schmidt had just obtained the spectrum of 3C9 in which for the first time one could see the Ly α emission line, and they were as a consequence able to apply the test and make people pay attention to it. For this reason I have called this the Gunn-Peterson test.

The opacity due to the Lyman α resonance is again given by equation (45), and even the statistical weight factor happens to be the same because it is a transition from $L = 0$ to $L = 1$. Here $h\nu_a / kT_s >> 1$, for otherwise a substantial fraction of the atoms would be in the 2P state, which is hard to arrange because the half life is $\sim 10^{-9}$ sec, so there is no correction for stimulated emission, and

$$\kappa_\nu = \frac{3}{8\pi} n\Lambda_a \, \delta(\nu - \nu_a) \, \lambda_a^2 .$$

Here $\lambda_a = 1215$ A is the resonance wavelength, and Λ_a is the transition rate (56). The optical depth (48) becomes

$$\tau = \frac{3}{8\pi} n(\mathrm{HI}) \, (1 + z_a)^3 \, \Lambda_a \, \lambda_a^3 \, H(z_a)^{-1}, \tag{60}$$

where z_a is the redshift at which the radiation passed through the resonance – if the wavelength observed is λ, the redshift is

$$1 + z_a = \lambda / 1215 \text{ A}.$$

The number density of atomic hydrogen at the epoch z_a has been written as $n(\mathrm{HI}) \, (1 + z_a)^3$, where $n(\mathrm{HI})$ would be the present density if the ionization were constant. $H(z_a)$ is the value of Hubble's constant at epoch z_a.

In the spectra of distant quasars where one can see the redshifted Ly α line one would look for a discontinuous decrease in brightness going shortward of the line, corresponding to the optical depth (60). No such effect is seen, so the optical depth must be below unity. This gives the upper limit

$$n(\mathrm{HI}) < \frac{8\pi}{3} \, \frac{H(z_a)}{(1 + z_a)^3 \, \Lambda_a \, \lambda_a^3} . \tag{61}$$

To establish orders of magnitude let us adopt an Einstein-deSitter model (eq. I-21). Then

$$H(z_\alpha) = (1+z_\alpha)^{3/2}\,H\,.$$

The quasars are observed at redshift $z \gtrsim 2$. Collecting, we have

$$n(HI) < 4.7 \times 10^{-12}\,h\,cm^{-3}, \qquad (62)$$

which miraculous limit is some seven orders of magnitude below the critical Einstein-deSitter density.

Bahcall and Salpeter pointed out that much the same limit should be possible for molecular hydrogen, for which the corresponding electronic transition is at a somewhat higher energy.[40] The observational limit of Field, Wampler and Solomon gave[41]

$$n(H_2) < 6 \times 10^{-10}\,cm^{-3}, \qquad (63)$$

where again the density is extrapolated to the present assuming $n \propto (1+z)^3$.

The limits (62) and (63) are of course irrelevant if the quasars are not at the distances suggested by the cosmological interpretation of their redshifts. This assumption is beginning to seem reasonable with the discovery, initiated through the efforts of Bahcall, that three objects that seem like quasars are associated in the sky with apparently ordinary galaxies at the same redshift.[42] The subject still is under discussion, however, so it would be well not to count too heavily on the limits (62) and (63).[43]

It might be mentioned finally that the limits (62) and (63) are extrapolated from redshift $z \gtrsim 2$, and it is of course another problem to say how the relative abundances of HI, HII and H_2 in fact have changed since then. Some elements of this question are discussed in Chapter VII.

iv) *Tests for an Intergalactic Plasma*

A second important line of thought on diffuse intergalactic matter is that there may be in intergalactic space a dilute more or less uniformly distributed plasma, free electrons and protons. The following is a list of the main proposals by which one might hope to detect an intergalactic plasma. Only two of the more popular tests will be discussed in detail here.

(a) *Free-free emission*: This is bremsstrahlung radiation by the electrons accelerated in the field of the ions. As will be described, there is some chance that a soft X-ray background consistent with this effect has been detected. The main problem with this test is that the expected effect is so very model dependent.

(b) *Free-free absorption*: This is just the reverse of (a), but one would look for it in the very long wavelength part of the spectrum of a radio source because opacity goes as λ^2 (eq. (65) below).[44,45]

(c) *Recombination radiation*:[37,44,46] Electrons and protons in the plasma will be recombining, sometimes to an excited state of hydrogen followed by radiative transitions to the ground state, through 2P-1S, producing among other things Lyman α radiation. Just as in the case of 21 cm emission this process would cause a step in brightness going longward of 1215 A. As described below this may ultimately prove to be the cleanest test for an intergalactic plasma, for the height of the step at 1215 A depends only on the present values of parameters.

(d) *Thomson Scattering*:[39,40,44,47] This produces a sort of general intergalactic fog. The brightness of a distant object is appreciably reduced (optical depth through the fog $\gtrsim$ 1) if $z \gtrsim 8$ and plasma density $\sim \rho_c$. This is a slippery test because the fog affects all wavelengths (with $h\nu \ll m_e c^2$) equally, so unless one can make some firm *a priori* statement on how bright a distant object should have been, one has no idea how much its observed brightness has been reduced by the fog.

(e) *Phase lag effect*:[48] The velocity of propagation of a radio signal in the plasma decreases with increasing wavelength, so if (as is observed) a distant quasar rapidly changed luminosity one would look for a delay in the observed brightness change at long wavelengths relative to short.

(f) *Faraday rotation*:[49] *If* there is an intergalactic plasma, *and if* there is a fairly homogeneous intergalactic magnetic field, one would look for a systematic Faraday rotation of the observed polarization of radio radiation from a distant polarized source.

Bremsstrahlung Radiation

The first of these tests will be discussed in some detail because it has had the greatest application, and, perhaps, success. We need the volume emission j_ν due to the bremsstrahlung radiation, and the easiest way to get it without actually computing anything is to consider first the time reversed process, free-free absorption.

Suppose there is in the plasma an applied electromagnetic wave $E = E_0 \cos(\omega t)$. The field is supposed to be intense, the number of photons per mode $\mathcal{N} \gg 1$, and it is assumed that $\hbar\omega/k\,T_e \ll 1$, where T_e is the plasma temperature, so that absorption of a single photon does not appreciably perturb the electron. Under these conditions classical electrodynamics applies. The velocity of a free electron is then

$$v = v_0 + \frac{eE_0}{m_e\omega} \sin \omega t,$$

where v_0 is the ordinary thermal velocity. The electron has mean energy

$$\frac{1}{2} m_e \langle v^2 \rangle = \frac{3}{2} k\,T_e + \frac{e^2 E_0{}^2}{4 m_e \omega^2}. \tag{64}$$

When the electron suffers a close collision with an ion the extra energy in the second term in (64) is thermalized, the energy coming from the electromagnetic field. What is the cross-section for such a "thermalizing" collision? When the electron-ion distance is r the electrostatic potential energy is e^2/r, and the velocity of the electron is seriously perturbed when this quantity becomes comparable to the electron kinetic energy $\frac{3}{2} k\,T_e$, that is, when the impact parameter $r \lesssim 2\,e^2/(3k\,T_e)$, so the cross-section is, in order of magnitude,

$$\sigma \sim \pi r^2 \sim \frac{4\pi}{9} \frac{e^4}{(k\,T_e)^2}.$$

The energy dissipated per unit volume is the collision rate $\sim \sigma\, n_e v_0$ multiplied by the energy transfer per collision, and this quantity is by definition the opacity κ multiplied by the energy flux in the applied electromagnetic field,

$$\frac{\kappa E_0^2 c}{8\pi} \sim \left(\frac{4\pi e^4}{9(k T_e)^2}\right) n_e{}^2 \left(\frac{3k T_e}{m_e}\right)^{1/2} \left(\frac{e^2 E_0^2}{4m_e \omega^2}\right),$$

or

$$\kappa \sim \frac{0.4 e^6 n_e{}^2}{(m_e k T_e)^{3/2} \nu^2 c}.$$

More careful computation gives[50]

$$\kappa = \frac{1.93 e^6 n_e{}^2}{(m_e k T_e)^{3/2} \nu^2 c}.$$

Still more careful computation gives

$$\kappa = \frac{1.93 e^6 n_e{}^2}{c\, m_e{}^{3/2} (k T_e)^{1/2} h\nu^3} (1 - e^{-h\nu/k T_e}). \tag{65}$$

The term in parentheses represents the correction to the opacity for stimulated emission (cf. eq. 45). It is not surprising that the simple-minded calculation given here gives the opacity in the limit of large correction for stimulated emission, because it was assumed to start with that $h\nu/k T_e \ll 1$.

In the usual way the radiation rate j_ν is, from (44) and (65),

$$j_\nu = \kappa_\nu P_\nu$$

$$= \frac{3.86 e^6 n_e{}^2 e^{-h\nu/k T_e}}{c^3 m_e{}^{3/2} (k T_e)^{1/2}} \tag{66}$$

$$= 5.4 \times 10^{-39} n_e{}^2 T_e{}^{-1/2} e^{-h\nu/k T_e} \text{ ergs cm}^{-3} \text{ sec}^{-1} \text{ Hz}^{-1} \text{ ster}^{-1}.$$

Equations (65) and (66) give reasonably good approximations to free-free absorption and emission in situations of interest. A list of correction factors to (65) and (66) is given by Karzas and Latter.[51]

The total bremsstrahlung emission rate is obtained by integrating (66) over frequency and multiplying by 4π. The result is

$$4\pi j = 1.44 \times 10^{-27} T_e{}^{1/2} n_e{}^2 \text{ ergs cm}^{-3} \text{ sec}. \tag{67}$$

This divided into the heat capacity $3\, n_e\, kT$ for the gas of free electrons and protons gives the characteristic cooling time used in the discussion of the Coma cluster.

The route by which eq. (66) came to be applied to an intergalactic plasma is as follows. In 1958 Gold and Hoyle[52] discussed the idea that, in the Steady State cosmology, the matter continuously created might be neutrons. The decay energy of the electrons would provide heat to make a hot intergalactic plasma. The neutron decay energy is 0.78 MeV, and on averaging over the electron energy distribution one finds that the mean kinetic energy given to the electron is

$$Q = 0.30 \text{ MeV},$$

the rest going to the neutrino. To get at the resulting plasma temperature consider the energy balance within a fixed (constant) spherical volume V, radius R. If n is the mean density (protons cm^{-3}) the rate of production of neutrons in V is $3\, n\, HV$ (eq. I-25), so the net rate of production of heat is $3n\, HVQ$. The volume is losing energy because material is streaming out at the speed HR, doing work against the external pressure $2n\, k\, T_e$, and carrying away material with internal energy density $3n\, k\, T_e$. The net rate of change of energy within V is then

$$\frac{d\mathcal{E}}{dt} = 3n\, HQ\, (\tfrac{4}{3}\pi\, R^3) - (5n\, k\, T_e)\, (4\pi\, R^2)\, HR,$$

which must vanish in the steady state, giving

$$\begin{aligned} T &= Q/5k \\ &= 60 \text{ keV} \\ &= 0.7 \times 10^9 \text{K}. \end{aligned}$$

In 1962 it was discovered that there is an isotropic extraterrestrial X-ray background.[53] Hoyle pointed out that, if there were a hot intergalactic plasma, as is perhaps expected in the Steady State cosmology, the bremsstrahlung emission by the plasma might account for this X-ray background.[54] However, more detailed computation by Gould and Burbidge[55] revealed that the expected X-ray flux from the hot plasma would be larger than is observed by a factor of $\sim$ 100 if the plasma density were equal to ρ_c. This was confirmed by Field and Henry.[56]

In the Steady State model it is easy to compute the expected spectrum of the X-ray background. Since self-absorption is unimportant, the brightness of the X-ray radiation satisfies (eq. 11)

$$\frac{d}{dt} i(t, \nu(t)) = -3\frac{\dot{a}}{a} i + j(t, \nu(t))\, c\,. \tag{68}$$

This equation applies in a homogeneous expanding cosmological model, evolving or Steady State. In the Steady State model $\dot{a}/a = H$ is a constant, i and j depend on time only through $\nu(t)$, so the solution to (68) is

$$i(\nu) = c \int_{-\infty}^{t} e^{-3H(t-t')}\, j(\nu e^{H(t-t')})\, dt'.$$

With equation (66) this yields

$$i(\nu) = 5.4 \times 10^{-39} c\, n_e^{\,2}\, T_e^{-\frac{1}{2}} \int_{-\infty}^{t} \exp\left[-3H(t-t') - \beta e^{H(t-t')}\right] dt',$$
$$\beta = h\nu/k\, T_e. \tag{69}$$

With the variable change $x = e^{H(t-t')}$, the integral in (69) becomes

$$H^{-1} \int_{1}^{\infty} dx\, e^{-\beta x}/x^4,$$

which when $\beta \ll 1$ reduces to $(3H)^{-1}$ and when $\beta \gg 1$ reduces to $H^{-1} e^{-\beta}/\beta$. That is, the spectrum is flat longward of a cutoff at $h\nu \sim k\, T_e$.

If the plasma density is equal to the Einstein-deSitter value (eq. 1), and if $T_e = 7 \times 10^8$ K, then (69) gives

$$i_\nu = 8 \times 10^{-26}\, h^3 \text{ ergs cm}^{-2} \text{ sec}^{-1} \text{ ster}^{-1} \text{ Hz}^{-1},$$
$$h\nu \lesssim 60 \text{ keV}. \tag{70}$$

The observed spectrum looks not at all like (69). Rather i_ν varies as $\nu^{-\alpha}$, with α approximately 0.7 up to 40 keV energy, and at higher energies $\alpha \sim 1.5$.[57] At 60 keV energy (70) exceeds the observed brightness by a factor of $20\,h^3$. That is, if the plasma temperature were 7×10^9 K the plasma density would have to be less than $\sim 0.2\,h^{-3/2}\,\rho_c$.

In an evolving cosmology the situation is much less easily summarized because the time dependence of T_e and the expansion parameter a are not known ahead of time. To get some numbers let us adopt the Einstein-deSitter cosmological model, $a(t) \propto t^{2/3}$, and let us assume that the plasma forms at the epoch fixed by redshift z_i and that the amount of plasma in a comoving volume has remained constant since then, so that the plasma density satisfies

$$n_e(t) = n(HII)\,(a_0/a(t))^3 .$$

One might suppose that, due to some heat source like intergalactic cosmic rays, the plasma temperature has been constant since z_i. With all these assumptions the solution to (68) for the present brightness of the X-ray background becomes

$$i_\nu = 5.4 \times 10^{-39}\,\frac{n(HII)^2\,c}{T_e^{1/2}\,H}\left(a_0^{3/2}\int_{a_i}^{a_0}\frac{da}{a^{5/2}}\,e^{-\beta\,a_0/a}\right), \qquad (71)$$

$$\beta = \frac{h\nu}{k\,T_e} = \frac{11.6\,E(keV)}{T_6},$$

where T_6 is the plasma temperature in units of 10^6 K. Of some current interest is the idea that the plasma might be cool, $T_e \lesssim 10^6$ K, in which case the best test is the soft X-ray flux because the hard end is cut off by the exponential factor. The X-ray spectrum has been observed down to ~ 0.25 keV, the flux at still lower energies (it is thought) being entirely absorbed by photo-ionization processes in the interstellar gas. If $T_6 \lesssim 1$ it is a reasonably good approximation to take $\beta >> 1$ even at $E = .25$ keV, in which case if $z_i >> 1$ (71) reduces to

$$i_\nu = 2.2 \times 10^{-24} \left(\frac{n(\mathrm{HII})}{n_c}\right)^2 h^3 \, T_6^{1/2} \, e^{-2.9/T_6} \quad \text{ergs cm}^{-2} \text{ sec}^{-1} \text{ ster}^{-1} \text{ Hz}^{-1}, \tag{72}$$

at 0.25 keV energy. There is some reason to believe that the soft X-ray background intensity is in fact larger than expected from a power law fitted to the spectrum above 1 keV energy, an effect first proposed by Henry, *et. al.*, in 1968,[58] and the thought is that this might indicate that a new source is dominating the soft X-ray background, the intensity at 0.25 keV being

$$i_\nu \sim 1 \times 10^{-24} \text{ ergs cm}^{-2} \text{ sec}^{-1} \text{ ster}^{-1} \text{ Hz}^{-1}. \tag{73}$$

This number is highly uncertain because of the question of the correction for absorption in the Galaxy, but the rough coincidence with the coefficient in the model result (72) is striking, and may conceivably be telling us that there is an intergalactic plasma with density $\sim \rho_c$ and temperature $\sim 10^6$ K. It will be noted, however, that in comparing (72) to (73) we are fitting a two-parameter model, n(HII) and T_6, to one number only. Worse, it is easy to add parameters, as in the following model.

A model considered by Field and Henry[56] is that the plasma formed at redshift z_i and since has adiabatically cooled, $T \propto a(t)^{-2}$ (eq. I-13). In this model the X-ray brightness is

$$i_\nu = \frac{5.4 \times 10^{-39} \, n(\mathrm{HII})^2 \, c}{H \, T_e^{1/2}} \int_{a_i}^{a_0} \frac{a_0^{1/2} \, da}{a^{3/2}} \, e^{-\beta a/a_0},$$

where β is defined as in equation (71). One possibility is that $\beta >> 1 + z_i$. In this case the largest contribution to the integral comes from epoch z_i, and the flux at .25 keV is approximately

$$i_\nu \sim 2.2 \times 10^{-24} \left(\frac{n(\mathrm{HII})}{n_c}\right)^2 h^3 \, T_6^{1/2} \, (1 + z_i)^{3/2} \, e^{-2.9(1+z_i)^{-1} \, T_6^{-1}}$$

$$\text{ergs cm}^{-2} \text{ sec}^{-1} \text{ ster}^{-1} \text{ Hz}^{-1}.$$

Here again we can choose parameters to fit (73), and it will be noted that, if we want to make z_i fairly large, n(HII) can be correspondingly reduced. Yet more detailed models have been computed by Weymann.[46,59]

It is apparent that, although the soft X-ray flux may be considered as suggestive of an intergalactic plasma, we cannot use the flux to find the plasma density because the expected flux is so sensitive to the model. Also, until we have a convincing explanation for the origin of the harder X-ray background, it will be hard to argue that the soft X-ray excess really is a new effect, not just a wiggle in an unknown source.

Recombination radiation

This test apparently was first applied by Field,[44] and has been considered most recently by Kurt and Sunyaev.[37] In principle one would like to have a measure of the step in intensity going longward across the local Lyman α line, thereby subtracting out any continuous background such as might originate in other possible intergalactic processes. The best we have at the moment are flux measurements just longward of 1215 A.

The rate of production of Lyman α photons per unit volume is

$$\alpha_{2P}(T_e)\, n_e^2,$$

where α_{2P} is the effective recombination coefficient taking account of recombination direct to 2P plus cascades from higher states. A table of these coefficients is given by Pengelly.[60] One finds in the standard way (eq. 41) that the radiation brightness resulting from Ly α recombination radiation looking just longward of 1215 A is

$$i_\nu = \alpha_{2P}(T_e)\, n(\mathrm{HII})^2\, hc/(4\pi H).$$

The limit on the ultraviolet background near 1215 A stated by Kurt and Sunyaev is

$$f \lesssim 2.5 \times 10^{-7} \text{ ergs cm}^{-2} \text{ sec}^{-1} \text{ ster}^{-1},$$

$$\Delta\lambda \sim 115 \text{ A},$$

which gives

$$n(\mathrm{HII}) \lesssim \left(\frac{4\pi\, H\lambda_\alpha^{\,2}\, f}{\alpha_{2P} hc^2 \Delta\lambda} \right)^{1/2} .$$

The very pleasant feature of this limit is that it depends only on local parameters, H and the plasma temperature T_e, and it is not a very sensitive function of T_e. For example, from Kurt and Sunyaev's flux limit and Pengelly's recombination coefficients one finds that

$$\begin{aligned} n(\mathrm{HII}) &\lesssim 3.5 \times 10^{-5}\, h^{1/2}\, \mathrm{cm}^{-3},\ T_e = 1 \times 10^4\ \mathrm{K}, \\ &\lesssim 6.8 \times 10^{-5}\, h^{1/2}\, \mathrm{cm}^{-3},\ T_e = 1 \times 10^6\ \mathrm{K}. \end{aligned} \tag{74}$$

v) *Intergalactic Dust*

People who like to confound cosmologists often mention what is variously called the meteorite problem, the planet problem, the football problem. Matter in such forms could add an enormous intergalactic mass density, yet defy detection. However, when realistically considered this does not seem to be a serious problem. As was remarked at the beginning of this chapter there is good evidence from the compositions of old stars that hydrogen is by far the most abundant element in original composition of the Galaxy, perhaps even that the Galaxy formed from material containing no atoms more massive than helium. Thus it seems unreasonable to look for appreciable amounts of condensed intergalactic non-volatile matter without expecting to encounter very much larger amounts of hydrogen. One could then counter (as in Sec. c-i) that the hydrogen might be condensed, intergalactic snow.[61] Here again we can place a lower bound on the possible size of the snow particles from the condition that the light from the most distant galaxies is not strongly attenuated. Let us consider spherical snow grains, radius r, density ρ_I, hence mass $4\pi\, \rho_I\, r^3/3$, and number density $n = 3\rho(t)/(4\pi\, \rho_I\, r^3)$, where

$$\rho(t) = \rho(S)\, a_0^3/a(t)^3 ,$$

$\rho(S)$ being the mean mass density provided by the snow grains now. In time interval dt photons from a distant galaxy go a distance cdt, and if r is greater than one micron the probability for scattering by the snow is

$$dP = 2\pi r^2 \cdot \frac{3\,\rho(t)}{4\pi\,\rho_I\, r^3} \cdot cdt\,,$$

so the brightness of a distant galaxy is reduced due to the scattering by the factor

$$i_0/i_i = e^{-\tau},\ \tau = \frac{3c}{2r}\int_{t_i}^{t_0} \rho(t)\, dt/\rho_I,$$

where the radiation observed now, at time t_0, left the galaxy at time t_i. To fix orders of magnitude let us adopt as usual the Einstein-deSitter cosmological model. This yields

$$\tau = \frac{c\,\rho(S)}{Hr\,\rho_I}\left[(1+z)^{3/2} - 1\right]. \tag{75}$$

The redshift-magnitude relation for bright galaxies shows no appreciable deviation from linearity (to 30 percent accuracy) to $z \cong 0.2$, so it would appear that $\tau(z = 0.2) \lesssim 1$. More detailed fits to the optical depth have been discussed by Dufay and by Nickerson and Partridge.[62]

The density within the grains would be $\rho_I \leq 0.07$ g cm^{-3}, the density of solid hydrogen. Collecting, we have

$$r \gtrsim 1.0h\,\rho(S)/\rho_c\ \text{cm}. \tag{76}$$

If $\rho(S)$ were comparable to the Einstein-deSitter density ρ_c, evidently the grains would have to be fairly large snowflakes.

The above limit is based on the assumption $r >> \lambda$, where λ is the wavelength of the light. In the other limiting case the grain might be approximated as a dielectric sphere, in which case the scattering probability is reduced by the factor $\sim (r/\lambda)^4$, and with it the optical depth (75). The limit here becomes

$$r << 100\ \text{A}\,(\rho_c/\rho(S))^{1/3}\, h^{-1/3}. \tag{77}$$

Now let us consider the stability of the grains. The grains are thought to be bathed in microwave radiation at temperature 2.7 K. The radiation spectrum may be blackbody or there may be a sub-millimeter excess beyond blackbody (Chapter V). Thus, in the absence of excess cooling by evaporation, the equilibrium grain temperature is at least 2.7 K. We need the equilibrium vapor pressure of solid H_2 at this temperature.[61] The pressure being impressively small we must extrapolate from higher temperature, and the convenient extrapolation formula is

$$P(T) = Ae^{-B/T},$$

with A and B constants. It will be recalled that this is based on the Clausius-Clapeyron relation and the assumptions (1) the latent heat for evaporation of H_2 (which comes off much more easily than H) is constant, (2) the equilibrium H_2 vapor obeys the ideal gas law. Borovish, *et. al.*,[63] have measured the vapor pressure in the temperature range 3.4 to 4.5 K, and their results are well fitted by the formula (based on a least squares fit to their data)

$$P(T) = e^{9.90 - 104/T} \text{ mm}, \tag{78}$$

the measured pressures being in the range of 1.8×10^{-6} to 1.0×10^{-9} mm. At $T = 10$ K this formula gives $P = 0.60$ mm, compared to the measured value 1.93 mm,[64] while at the triple point, $T = 13.80$ K, (78) gives $P = 10.6$ mm, compared to the measured value 53 mm. The extrapolation toward lower temperature should be more reliable than the extrapolation to higher because the latent heat should be more nearly constant and the ideal gas law should be a better approximation. There is good reason therefore to believe that (78) may be extrapolated to 2.7 K. The result is

$$\begin{aligned} P_e\,(2.7\text{ K}) &= 4 \times 10^{-13} \text{ mm}, \\ n_e\,(2.7) &= 1.3 \times 10^{6} \text{ cm}^{-3}. \end{aligned} \tag{79}$$

The result is some eleven orders of magnitude larger than the estimate of Hoyle, Reddish, and Wickramasinghe.[61] Apparently, this is because they extrapolated from a higher temperature and they assumed a higher effective latent heat.

It is apparent from (79) that intergalactic hydrogen grains in H_2 gas density $\lesssim 10^{-5}$ cm^{-3} would be unstable against evaporation at 2.7 K, and would be unstable unless the temperature were 1.6 K or lower. To estimate the evaporation rate let us suppose the grain temperature is 2.7 K. Then if the grains were in an H_2 gas with density equal to (79), also at 2.7 K, the evaporation rate would be equal to the rate of capture of H_2 molecules by the grains. The latter may be estimated by making the reasonable assumption that every molecule that hits the grain sticks. On this argument the evaporation rate is

$$\frac{dN}{dt} = 4\pi r^2 n_e \left(\frac{3kT}{m_2}\right)^{1/2} ,$$

where m_2 is the mass of a hydrogen molecule. A characteristic evaporation time is dN/dt divided into the number of molecules in the grain,

$$t_{evap} = N/\dot{N} \sim 3 \times 10^{11} \text{ r sec} , \tag{80}$$

where r is the grain radius measured in cm.

As the grain evaporates, it cools, and absorbs heat from the microwave background. The important quantity is the ratio of heat loss rate by evaporation to heat gain from the ambient radiation. Assuming the grains are black and have radius $\gtrsim$ 1 cm this ratio is

$$\frac{\ell \dot{N}}{\pi r^2 aT^4 c} = 0.12 . \tag{81}$$

Here ℓ is the latent heat of sublimation per molecule, and a is the Stefan-Boltzmann constant. Because this ratio is comparable to unity the grains will be slightly cooler than the radiation equilibrium and the evaporation rate correspondingly reduced, but not by a very large factor. Thus the grains can last an expansion time, $\sim 3 \times 10^{17}$ sec, only if the grain radius $\gtrsim 10^6$ cm = 10 km.

The ratio (81) was based on the assumption that r $\gtrsim$ 1 cm. If 1 cm > r $\gtrsim$ 100 A the rate of absorption of radiation is very much reduced, but it may be verified that the grains still ought to disappear in 10^{17} sec.

If r were any smaller than 100 A the curvature effect should substantially increase the equilibrium vapor pressure and the evaporation rate.

It is concluded that intergalactic hydrogen could exist in solid form as a reasonably permanent feature of intergalactic space only if it were in rather substantial lumps, radius $\gtrsim$ 10 km. Of course this does not prove that the lumps do not exist, but it is hard to see how they could have formed.

e) *Relativistic Matter*

According to general relativity theory we must include in the mass density ρ in equation (I-10) the radiation energy density summed over all frequencies and divided by the square of the velocity of light. Thanks to the dramatic progress in radio astronomy and space science we now know the extraterrestrial electromagnetic radiation background over most of the frequency range from 1 Mc/sec = 10^6 Hz on the radio side to 10^{22} Hz on the γ ray side. The major gap in this band is in the infrared, but even here the upper limits are improving.[65] A minor gap is in the range $\sim$ 50 A to 912 A, where the Galaxy is opaque due to photo-ionization of atomic hydrogen.

The desired quantity is the radiation brightness i_ν integrated over all frequencies. Because the observed frequencies span some 15 orders of magnitude it is very convenient to rewrite this integral as

$$\rho\,(\text{radiation}) = \frac{4\pi}{c^3} \int_{-\infty}^{\infty} \nu i(\nu)\, d \ln \nu \,. \tag{82}$$

Thus if νi_ν is plotted as a function of $\ln \nu$ the area under the curve is directly proportional to the equivalent mass density contributed by the radiation.

The observational situation is shown in Figure IV-2.[66] The quantity shown on the vertical scale is

$$\nu \rho_\nu \equiv \frac{4\pi}{c^3} \nu i_\nu \,.$$

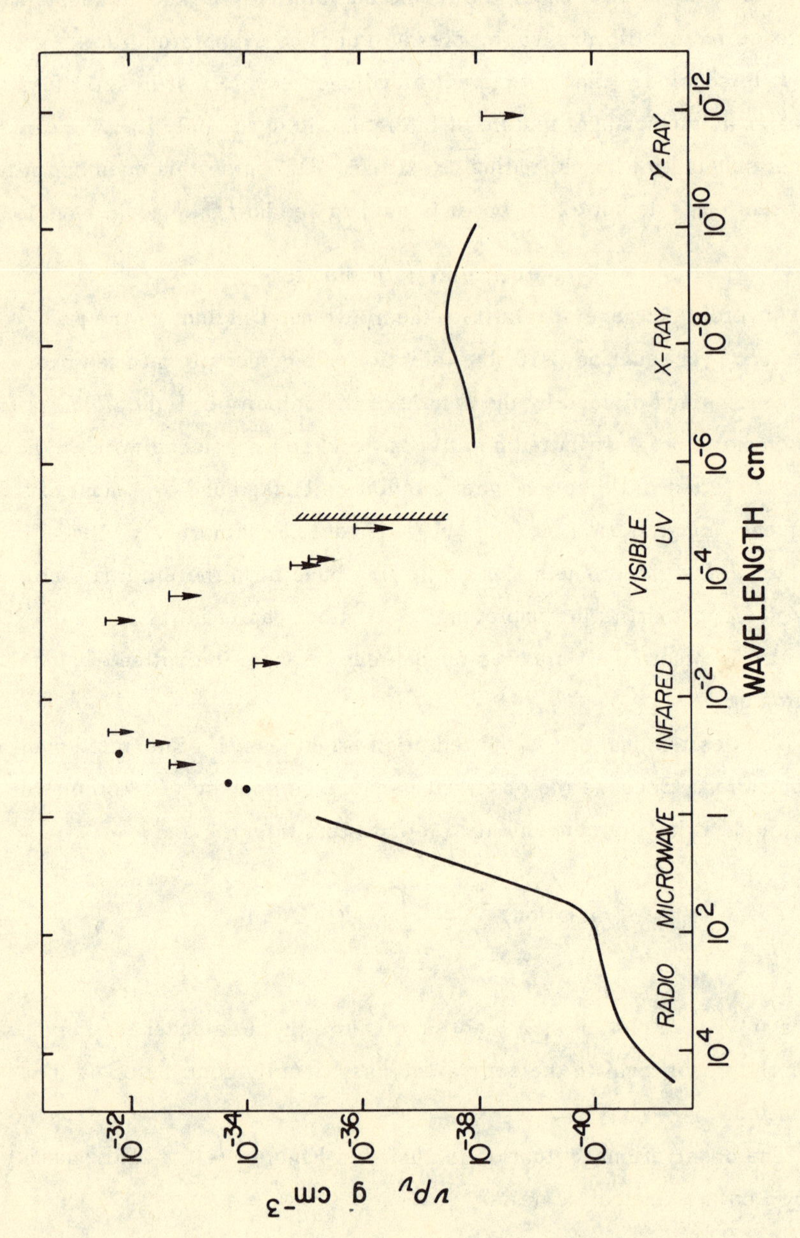

Fig. IV-2. Cosmic Electromagnetic Radiation.

It has been plotted on a logarithmic scale, so to estimate the mass density one must imagine stretching the vertical scale to a linear one in $\nu \rho_\nu$, then estimate the area under the curve.

With the exception of the longest wavelength points the radio and microwave spectrum up to 3 mm wavelength has been measured from the ground. It will be noted that there is an abrupt change in slope at $\lambda \sim 75$ cm, the brightness at shorter wavelength proportional to $i_\nu \propto \nu^2$ to good accuracy. As will be described in Chapter V there is some chance that this is blackbody radiation left over from the Big Bang. The infrared part of the spectrum from 3 mm to 1 micron must be measured above the atmosphere because the atmosphere is radiating so strongly. Lines with arrows represent upper limits. The apparently contradictory situation around 500 microns is discussed in detail in Chapter V. The upper limit in the middle of the infrared is a recent result of McNutt and Feldman,[67] and the upper limits in the 1-10 micron range were obtained by Harwit, *et. al.*[68]

The two upper limits in the visible spectrum were mentioned in Section 6 of this Chapter,[8,9] the ultraviolet limit of Kurt and Sunyaev was mentioned in Section d-iv,[37] the diffuse X-ray background is taken from a summary by Schwartz, *et. al.*,[57] and the last point represents an upper limit or possible detection of an isotropic γ-ray background at energy $\geq$ 100 MeV.[69]

It is apparent from the Figure that the observed part of the radio spectrum flux, the optical and ultraviolet flux, and the X-ray and γ-ray flux all amount to a small fraction of $\rho_c = 10^{-29}$ g cm^{-3}. The infrared part may contain the largest fraction of the total, and there are still serious gaps here, but it appears that unless there is a very substantial infrared background concentrated in bands at the right wavelengths the integrated electromagnetic radiation energy density is substantially less than $\rho_c c^2$. Infrared astronomy is just beginning, and we can expect to have much better information in the next few years.

There remains the question of energy flux outside the frequency range of Figure 1. Gould and Schréder[70] pointed out that the lifetime of a very high energy photon is sharply limited by electron pair production on microwave background photons. Let us estimate the magnitude of this effect on the assumption that the microwave radiation is part of a Planck thermal spectrum at temperature $T_0 = 2.7$ K. Then the typical energy of a photon is $q_0 \sim kT_0$, and the threshold energy q of a γ-ray photon for electron pair production is

$$q = \frac{m_e^2 c^4}{q_0} \sim 10^{15} \text{ eV}.$$

Near threshold the cross-section for pair production is $\sim \sigma_T$, the Thomson cross-section. The number density of thermal photons is roughly the blackbody energy density aT_0^4 divided by q_0,

$$n_\gamma \sim aT_0^3/k \sim 10^3 \text{ cm}^{-3}.$$

The mean free time for a γ-ray near threshold is then

$$t_p \sim (\sigma_T n_\gamma c)^{-1} \sim 10^{11} \text{ sec},$$

very much less than H^{-1}. A more detailed integration over the blackbody energy distribution shows that the mean life of a photon is less than H^{-1} if $q > 10^{14}$ eV.[70] We could assume that sources are providing an interesting density in energetic photons beyond this limit over the characteristic time H^{-1} only if we were willing to accept a much greater density in degraded energy.

A photon with energy $\sim 10^{14}$ eV would cause a "muon poor" extensive air shower, from which we have an upper limit on the very energetic γ-ray flux. A limit is[71]

$$f \lesssim 2 \times 10^{-11} \text{ photons cm}^{-2} \text{ sec}^{-1} \text{ ster}^{-1},$$
$$q \gtrsim 3 \times 10^{13} \text{ eV},$$

which corresponds to

$$\rho\,(q > 3 \times 10^{13} \text{ eV}) \lesssim 10^{-40} \text{ g cm}^{-3}.$$

At the other end of Figure IV-2 the interesting limit is an intergalactic magnetic field. If the typical field is $10^{-6} B_6$ Gauss the equivalent mass density is

$$\rho\,(B) = \frac{B^2}{8\pi\, c^2} = 4.4 \times 10^{-35}\, B_6^{\,2}\, g\, cm^{-3}.$$

The magnetic field in the Galaxy is $B_6 \sim 1$, and if the intergalactic magnetic field is not greater than this the contribution to the mass density is negligible.

Cosmic rays might be listed in this section. The energy density of relativistic particles in the Galaxy is put at ~ 1 eV cm^{-3}, or $\sim 10^{-33}$ g cm^{-3}. We do not know what fraction of this might be only local, what part extragalactic.

Neutrinos might also be mentioned, but will not be discussed because the observational limits are so poor. There is ample room for missing mass $>> 10^{-29}$ g cm^{-3} for the Universe in neutrinos anywhere in the energy range from 0.01 eV (where they would start to become degenerate) to 10^7 eV (where they would start to be manifest in deep mine neutrino experiments). Reviews are given by Ruderman and de Graff.[72]

Finally, as in the discussion of the Coma cluster, we must recognize that black holes could provide the missing mass, and, if they were not too large, there would appear to be no feasible direct way to detect them. If the black holes resulted from evolution of non-relativistic systems, perhaps galaxies of some sort, then we might expect that the collapse was attended by vigorous radiation which we might hope to detect.

f) *Summary*

The greatest obstacle to thinking that the usual estimate of the mass density provided by galaxies may have any meaning is the inability to make consistent pictures of the dynamics of groups and clusters of galaxies. This problem may mean that the conventional picture misses an important new physical effect, like spontaneous creation of galaxies, but it is entirely possible also that it only reflects our deep ignorance of the physical conditions, the parameters, in the problem. An important test will be the rich compact clusters of galaxies, for the regular forms of these systems strongly suggest they are stable and well mixed, and it should be possible to accumulate enough data to make precise application of the virial theorem. It is important also that, if the discrepancy persists, we can make rather detailed investigations on possible forms of hidden mass in these compact systems.

A second cause for worry is the observed range of mass-to-light ratios of galaxies, from unity or even less for irregulars and some spirals up to perhaps 100 for giant elliptical galaxies. The immediate cause of this large range is that the spirals and irregulars still have gas from which bright massive young stars are forming, while the ellipticals seem to have invested all their matter in small, dim, long-lived stars. We do not know why there should be this difference among galaxies. The worry is that the range of mass-to-light ratios could be even higher, that there could be dead galaxies with "normal" mass and "abnormally" low luminosity, and such objects would be hard to detect. A sort of variant of the "dead galaxies" problem is the thought that a galaxy could evolve into a black hole by contracting through its Schwarzschild radius. This should be the ultimate fate of an undisturbed galaxy, but the relaxation time for such extreme contraction is thought to be enormously greater than 10^{10} y. One might wonder, however, whether galaxies might not find themselves in a position where relaxation effects that people have not yet thought of greatly hasten the process.

It has frequently been remarked that if the typical correction factor to the mass needed to satisfy the virial theorem were applied directly to the estimated mean mass density of galaxies it would bring this quantity close to ρ_c. For example, in the case of the Local Group it would suffice to suppose that the two galaxies have massive faint extensive halos. However, this is not adequate for compact groups, where apparently we cannot explain (1) how the spirals in some groups have avoided disruption for 10^{10} y, and (2) how the systems avoided relaxing to a coalesced nucleus with the remaining objects an extended halo.[30]

Apart from the stability problem the main task clearly is to devise tests that might reveal each form of matter that contributes an interesting fraction of the total, so that we may in time arrive at a picture of the contents of the Universe. It would be naive to suppose that the mean mass density of the Universe may be estimated from the one component we can study in some detail, the galaxies, even if we could make sense of these objects, but on the other hand it would be equally unreasonable to conclude that we cannot hope to determine ρ. We have a number of local and global tests for diverse forms of matter, and there is no reason to think that our ingenuity has been exhausted. In the same spirit it would be too strong to say that conventional physics and the conventional view of the Universe are suffering a crisis because of the stability problem, because we are too ignorant to be sure we are not interpreting a perfectly "ordinary" phenomenon in a foolish way. We are still engaged in the simple task of accumulating the elementary observational picture of what the Universe is like.

REFERENCES

1. J. A. Oort, Solvay Conference on *Structure and Evolution of the Universe,* 1958, p. 163.
2. A recent summary is given by E. M. Burbidge and G. R. Burbidge, to be published in "*Galaxies and the Universe,*" vol. 9 of the series *Stars and Stellar Systems.*
3. T. Page, *Ap. J.* 136, 685, 1962.
4. F. Zwicky, *Ap. J.* 143, 192, 1966.
5. H. Arp, *Ap. J.* 142, 402, 1965.
6. P. J. E. Peebles and R. B. Partridge, *Ap. J.* 148, 713, 1967.
7. G. C. McVittie and S. P. Wyatt, *Ap. J.* 130, 1, 1959; G. J. Whitrow and B. D. Yallop, *M. N.* 130, 31, 1965; J. E. Felten, *Ap. J.* 144, 241, 1966; Z. Horak, *Astronautica Acta* 13, No. 1, 1967.
8. J. B. Oke and A. Sandage, *Ap. J.* 154, 21, 1968; A. D. Code, G. A. Welch and T. L. Page, paper presented at the OAO Symposium, Amherst, August 1971.
9. F. E. Roach and L. L. Smith, *Geophys. J. R. Astr. Soc.* 15, 227, 1968; C. F. Lillie, *An Empirical Determination of the Interstellar Radiation Field,* Ph. D. Thesis, University of Wisconsin.

10. F. Zwicky, *Helv. Phys. Acta.* 6, 110, 1933; S. Smith, *Ap. J.* 83, 23, 1936.
11. M. Schwarzschild, *A. J.* 59, 273, 1954.
12. G. C. Omer, T. L. Page and A. G. Wilson, *A. J.* 70, 440, 1965.
13. P. J. E. Peebles, *A. J.* 75, 13, 1970.
14. C. A. Muller, *B. A. N.* 14, 339, 1959.
15. N. J. Woolf, *Ap. J.* 148, 287, 1967; B. E. Turnrose and H. J. Rood, *Ap. J.* 159, 773, 1970.
16. H. Gursky, *et. al.*, *Ap. J.* 167, L81, 1971.
17. V. C. Reddish, *Q. J. Roy. Astro. Soc.* 9, 409, 1968.
18. R. Ruffini and J. A. Wheeler, *Significance of Space Research for Fundamental Physics*, ESRO, Paris, 1971.
19. S. van den Bergh, *Nature* 224, 891, 1969.
20. H. J. Rood and W. A. Baum, *A. J.* 73, 442, 1968.
21. M. S. Roberts, to be published in "*Galaxies and the Universe*," vol. 9 of the series *Stars and Stellar Systems*.
22. M. Demoulin, *Ap. J.* 157, 81, 1969.
23. F. D. Kahn and L. Woltjer, *Ap. J.* 130, 705, 1959.
24. J. Oort, *Astron. and Astrophys.* 7, 381, 1970.
25. H. Spinrad, *et. al.*, *Ap. J.* 163, L25, 1971.
26. C. K. Seyfert, *P.A.S.P.* 63, 72, 1951.
27. E. M. Burbidge and W. L. W. Sargent, Paper presented at Semaine d'Etude on Nuclei of Galaxies, Pontifical Academy of Sciences, Vatican City, April, 1970.
28. L. Spitzer and W. Baade, *Ap. J.* 113, 413, 1951.
29. E. M. Burbidge and G. R. Burbidge, *Ap. J.* 134, 244, 1961; W. L. W. Sargent, *Ap. J.* 153, 135, 1968.
30. V. A. Ambartsumian, *A. J.* 66, 536, 1961.
31. eg. G. de Vaucouleurs, *Ap. J.* 130, 718, 1959; G. R. Burbidge and E. M. Burbidge, *Ap. J.* 130, 629, 1959.
32. P. J. E. Peebles, *Ap. J.* 157, 45, 1969.
33. S. J. Goldstein, *Ap. J.* 138, 978, 1963. Earlier references are given in this paper.

34. A. A. Penzias and R. W. Wilson, *Ap. J.* 156, 799, 1969.
35. G. B. Field, *Ap. J.* 129, 525 and 536, 1959; 135, 684, 1962.
36. A. A. Penzias and E. H. Scott, *Ap. J.* 153, L7, 1968; R. J. Allen, *Astron. and Astrophys.* 3, 382, 1969.
37. V. G. Kurt and R. A. Sunyaev, *Zh. E.T.F. Pis'ma* 5, 299, 1967; Engl. tr. in *J.E.T.P. Letters* 5, 246, 1967; G. E. Thomas and R. F. Krassa, *Astronomy and Astrophysics* 11, 218, 1971; C. Lillie, paper presented at the OAO Symposium, Amherst, August 1971.
38. I. S. Shklovoski, *Astron. Zh.* 41, 408, 1964; Engl. tr. in *Soviet Astron.–A. J.* 8, 638, 1965; P. A. G. Scheuer, *Nature* 207, 963, 1965.
39. J. E. Gunn and B. A. Peterson, *Ap. J.* 142, 1633, 1965.
40. J. N. Bahcall and E. E. Salpeter, *Ap. J.* 142, 1677, 1965.
41. G. B. Field, P. M. Solomon and E. J. Wampler, *Ap. J.* 145, 351, 1966.
42. J. N. Bahcall, M. Schmidt and J. E. Gunn, *Ap. J.* 157, L77, 1969; J. E. Gunn, *Ap. J.* 164, L113, 1971.
43. eg. H. Arp, *Ap. J.* 162, 811, 1970.
44. G. Field, *Ap. J.* 129, 536, 1959.
45. W. C. Erichson and W. M. Cronyn, *Ap. J.* 142, 1156, 1965; P. D. Noerdlinger, *Ap. J.* 157, 495, 1969.
46. R. Weymann, *Ap. J.* 147, 887, 1967.
47. J. N. Bahcall and R. M. May, *Ap. J.* 152, 37, 1968.
48. F. T. Haddock and D. W. Sciama, *Phys. Rev. Letters* 14, 1007, 1965.
49. Y. Sofue, M. Fujimoto and K. Kawabata, *Publ. Astron. Soc. Japan* 20, 388, 1968; K. Brecher and G. R. Blumenthal, *Astrophysical Letters* 6, 169, 1970; R. G. Conway and J. A. Gilbert, *Nature* 226, 337, 1970.
50. C. W. Allen, *Astrophysical Quantities*, 2nd ed., (1963), p. 99.
51. W. J. Karzas and R. Latter, *Ap. J. Suppl.* 6, 167, 1961.
52. T. Gold and F. Hoyle, Paris Symposium on Radio Astronomy, 1958.
53. R. Giacconi, H. Gursky, F. R. Paolini and B. B. Rossi, *Phys. Rev. Letters* 9, 439, 1962.
54. F. Hoyle, *Ap. J.* 137, 993, 1963.
55. R. Gould and G. Burbidge, *Ap. J.* 138, 969, 1963.
56. G. Field and R. Henry, *Ap. J.* 140, 1002, 1964.

57. J. Silk, *Space Science Reviews* 11, 671, 1970; D. A. Schwartz, H. S. Hudson, and L. E. Peterson, *Ap. J.* 162, 431, 1970.

58. R. C. Henry, G. Fritz, J. R. Meekins, H. Friedman, and E. T. Byram, *Ap. J.* 153, L11, 1968; P. G. Shukla and B. G. Wilson, *Ap. J.* 164, 265, 1971 and earlier references therein.

59. R. Weymann, *Ap. J.* 145, 560, 1966.

60. R. M. Pengelly, *M.N.* 127, 145, 1964.

61. N. C. Wickramasinghe and V. C. Reddish, *Nature* 218, 661, 1968; F. Hoyle, V. C. Reddish and N. C. Wickramasinghe, *Nature* 218, 1124, 1968.

62. J. Dufay, *Galactic Nebulae and Interstellar Matter* (Philosophical Library) 1957, Chapter 19; B. Nickerson and R. B. Partridge, *Ap. J.* 169, 203, 1971.

63. E. S. Borovish, S. F. Grishin and E. Ya. Grishina, *Zh. Tekh. Fiz.* 30, 539, 1960.

64. H. J. Hoge and R. D. Arnold, *J. Res. NBS*, 47, 63, 1951.

65. P. J. E. Peebles, *Comments on Astrophysics and Space Physics*, 3, 20, 1971.

66. An earlier version is shown in P. J. E. Peebles, *Phil. Trans.* A264, 279, 1969.

67. D. P. McNutt and P. D. Feldman, *J.G.R.* 74, 4791, 1969, and to be published.

68. M. Harwit, *et. al.*, *A. J.* 71, 1026, 1966.

69. G. W. Clark, G. P. Garmire and W. L. Kraushaar, *Ap. J.* 153, L203, 1968.

70. R. Gould and G. Schréder, *Phys. Rev. Letters* 16, 252, 1966.

71. K. Kamata, *et. al., Can. J. Phys.* 46, S72, 1968.

72. M. Ruderman, *Reports on Progress in Physics* 28, 411, 1965; T. de Graff, *Astronomy and Astrophysics* 5, 335, 1970.

V. THE MICROWAVE BACKGROUND AND THE PRIMEVAL FIREBALL HYPOTHESIS

A Primeval Fireball is thermal blackbody radiation, a remnant from a presumed earlier epoch when the Universe was much denser and hotter than it is now. As will be described in this chapter there is strong evidence that a Primeval Fireball has been discovered. If this is substantiated by further measurements the great and profound significance will be that we will have direct evidence that the Universe did expand away from a "Big Bang," and we will know something about the temperature history in the Big Bang.

a) *Concept, Prediction and Discovery*

It will be assumed here that the Universe is homogeneous and isotropic, and that the Universe is expanding and evolving as suggested by Hubble's law, so that the younger Universe was more dense than it is now. It is assumed also that the matter in the younger Universe was hot, and that it had relaxed to thermal equilibrium, so that space was filled with blackbody radiation, which is the Primeval Fireball. To see what happens to the blackbody radiation as the Universe expands let us assume for the moment that the interaction of the radiation with matter may be neglected. The interaction will be discussed in Chapter VII, and it will be seen that, except for some extreme situations, the effect of the matter is expected to be very small. If absorption and emission by matter may be neglected, then by equation (IV-11) the brightness of the radiation satisfies

$$\frac{d}{dt}\left[a(t)^3\ i(t,\ \nu(t))\right] = 0, \qquad (1)$$

where it will be recalled that $\nu(t) \propto a(t)^{-1}$ (eq. I-15). If the radiation brightness at epoch t_i is given to be $i(t_i, \nu)$, then equation (1) says that at any later time t the brightness is

$$i(t, \nu) = \left(\frac{a(t_i)}{a(t)}\right)^3 i(t_i, \nu a(t)/a(t_i)). \tag{2}$$

We are assuming that the initial brightness is that of blackbody radiation at temperature T_i, say,

$$i(t_i, \nu) = \frac{2h\,\nu^3}{c^2}\left(e^{h\nu/k\,T_i} - 1\right)^{-1}, \tag{3}$$

and this equation in (2) gives

$$i(t, \nu) = \frac{2h\,\nu^3}{c^2}\left(e^{h\nu/k\,T(t)} - 1\right)^{-1},$$
$$T(t) \equiv T_i\, a(t_i)/a(t). \tag{4}$$

This equation says that the radiation spectrum evaluated at a later time t is still blackbody, but that the temperature is reduced by the factor $a(t_i)/a(t)$. This is the main point. If the assumptions leading up to (4) are valid a Primeval Fireball would be identified as such by the distinctive blackbody spectrum.

Another way to understand (4) is to imagine that some region of space is enclosed by a container with perfectly reflecting walls inside and out, and size $<< cH^{-1}$. The cavity thus formed is assumed to be expanding with the Universe, so the volume of the cavity $\propto a(t)^3$. It might be objected that the box alters the situation in the Universe, but on the average this is not so – the walls just interchange by reflection photons that would have passed through, and the expansion of the box assures symmetry of radiation incident on either side of the wall. Now by ordinary thermodynamics we know that the expansion cools the radiation in the cavity. Explicitly, Planck's formula says that the number of photons per mode in the cavity is (eq. IV-51)

$$\mathcal{N} = \left(e^{h\nu/kT} - 1\right)^{-1}.$$

The wavelength $\lambda = c/\nu$ of the mode varies in proportion to the cavity size $\propto a(t)$, which is the cosmological redshift (eq. I-15). Since the fractional change in ν per period ν^{-1} is small the expansion is adiabatic, so $\mathfrak{N}$ is conserved, and as $\nu \propto a(t)^{-1}$ the parameter T must vary as $a(t)^{-1}$, as before.

From the local point of view we have in this derivation a sensible energy conservation law – the radiation energy within the cavity is decreasing because the number of photons in the cavity is conserved and the energy per photon decreasing, and this energy evidently is going into the radiation pressure work against the expanding cavity. There is not a global energy conservation law, which should not be considered too distressing because general relativity does not admit an energy conservation law of the conventional sort. There is conservation of entropy, because entropy is a scalar quantity. The entropy per unit volume for blackbody radiation at temperature T is

$$S = \int_0^T \frac{dE}{T} = \frac{4}{3} aT^3,$$

where the radiation energy density is

$$E = aT^4, \tag{5}$$

and a is the Stefan-Boltzmann constant, not to be confused with the expansion parameter. If the expansion of the Universe is thermodynamically reversable entropy is conserved, and the entropy per unit volume must vary as $a(t)^{-3}$ because the volume over which the entropy is distributed varies as $a(t)^3$. In other words the entropy of the radiation per nucleon number (also a conserved number),

$$\frac{S}{n} = \frac{4aT^3}{3n}, \tag{6}$$

is a constant. It will be noted, however, that one does not need the traditional grain of dust to keep the radiation thermal. That is, equation (4) should not be considered a result of thermodynamics, relativistic or otherwise. It is an accident of the form of the relation of energy and momentum, such that energy may be scaled from momentum. The whole calculation can be repeated for a collisionless gas of non-relativistic particles, and one finds that, if the velocity distribution function initially is the Maxwell-Boltzmann form at temperature T_i, the distribution function again preserves a thermal spectrum as the Universe expands but the temperature drops as $T \propto a(t)^{-2}$. For a mildly relativistic gas, with temperature $T \sim mc^2/k$ (particle rest mass m), one finds that the initial thermal distribution function is *not* preserved by the expansion (in the absence of thermalizing collisions).

The speculation is that our Universe contains a Primeval Fireball at temperature $T_0 = 2.7$ K. The mass equivalent of this energy density would be (eq. 5)

$$\rho_\gamma = \frac{aT_0^4}{c^2} = 4.5 \times 10^{-34} \text{ g cm}^{-3},$$

five orders of magnitude less than the Einstein-deSitter density (IV-1). The number density of blackbody photons is found by multiplying (3) by $4\pi/(h\nu\, c)$ and integrating over ν,

$$\begin{aligned} n_\gamma &= \frac{8\pi}{c^3} \int \frac{\nu^2\, d\nu}{e^{h\nu/k T_0} - 1} \\ &= 60.4\, (k\, T_0/hc)^3 \\ &= 400 \text{ cm}^{-3}, \end{aligned} \tag{7}$$

if $T_0 = 2.7$ K. The ratio of photon density to baryon density in the Einstein-deSitter model would be

$$n_\gamma/n_c = 3.5 \times 10^7\, h^{-2}.$$

This is also close to the entropy per baryon (6) divided by Boltzmann's constant, S/nk, because $a \sim k^4/(hc)^3$.

The ratio of the heat capacity of matter and radiation would be, in rough order of magnitude,

$$n_c\, k/aT_0^3 \sim 10^{-8}.$$

The very small size of this number has the important consequence that if matter and radiation interact to any extent the matter should come to the radiation temperature. Apparently, we have two simple possibilities. If the matter has negligibly weak interaction with the radiation then the above analysis applies, the radiation cools as $a(t)^{-1}$, and of course preserves its thermal spectrum. If the interaction is strong enough that the matter absorbs an appreciable amount of radiation energy then the matter temperature should approach the radiation temperature, and we know from ordinary thermodynamics that in this case the matter cannot perturb the thermal form of the radiation spectrum whatever the interaction. There are of course ways to get around this, as discussed in Chapter VII.

The history of the discovery of the candidate for the Primeval Fireball radiation has been described in detail,[1] and only a few essential points need be repeated here. The behavior of radiation in an expanding Universe was first studied by Tolman, in 1934 and earlier, and a derivation of equation (4) is given in his book.[2] Tolman's result was applied by Gamow and Alpher in 1948 to their theory of the production of elements in the early Universe.[3] An earlier picture had been that the relative abundances of the elements were determined at some time in the past when all the matter in the Universe was concentrated in a dense hot super-star of some sort. The thought was that if the density and temperature were properly chosen the thermal equilibrium distribution of atomic masses might coincide with the observed abundance distribution, and this was indeed the case if one considered a restricted range of atomic masses. Gamow[4] criticized this picture on two grounds. First, the over-all form of the abundance curve, from lightest to heaviest nuclei, is not at all like the form expected from thermal equilibrium – if the parameters are adjusted to fit the light end the heavy elements are far more abundant than the model would predict. Second, the assumption of thermal equilibrium is questionable if the Lemaître cosmology is valid. If we trace the expansion of the Universe back to a state of high density then we find from equation (I-10) that the large ρ makes the expansion rate $\dot{a}/a$ large. That is, the element formation would have to be a dynamic process, not a question of statistical equilibrium. Gamow and Alpher proposed that the elements were built up in the early dense Universe by successive neutron capture at a temperature $kT \sim 1$ MeV. The temperature and density conditions they needed to make the scheme work may be used with equations (4) and (I-14) to find the radiation temperature at the present epoch, density $\sim 10^{-6}$ protons cm^{-3}. The result is $T_0 \sim 5K$, which may be compared to the observed microwave radiation intensity, which is consistent with blackbody radiation at temperature $T_0 = 2.7$ K.

Gamow gave a nice explanation of how the temperature is fixed in their theory.[5] One imagines that the original primeval matter was in the simplest form, free neutrons, protons and electrons. The neutrons and protons can react to form deuterium,

$$n + p \rightarrow d + \gamma, \tag{8}$$

but until the radiation has cooled to 10^9 K photo-dissociation breaks up the deuterium as fast as it forms. Once the Universe has expanded and cooled to the temperature $T \sim 10^9$ K the deuterium can accumulate and burn to build up heavier elements. The reaction (8) is the key to this process, so Gamow observed that he could estimate the matter density at temperature 10^9 K from the condition that the net probability of the reaction should be appreciable, so there will be produced an appreciable abundance of elements heavier than hydrogen, but not excessive, so that hydrogen will remain the most abundant element, as is observed. That is, he required

$$\sigma n(t)\, vt \sim 1, \tag{9}$$

where σ is the cross-section for radiative capture of neutrons by protons, n is the nucleon number density, v is a mean thermal velocity, and t is the age of the cosmological model measured from the singular point $t = 0$,

$\rho \to \infty$; and t is of course also a characteristic time scale for expansion of the Universe. All of these quantities are supposed to be evaluated at $T = 10^9$ K, when element building can commence. It is assumed in this equation that there are about equal numbers of neutrons and protons.

The time t is given directly by equation (I-3). Because the radiation mass density $\propto T^4 \propto a(t)^{-4}$, it dominates the matter mass density ($\propto a(t)^{-3}$) and the other terms on the right hand side of equation (I-3) when a is very small. The result is that when $T = 10^9$ K the expansion rate is just

$$\left(\frac{\dot{a}}{a}\right)^2 = \frac{8\pi\, GaT^4}{3c^2},$$

and the solution is

$$t = \left(\frac{3c^2}{32\pi\, GaT^4}\right)^{1/2},$$

which gives $t \sim 200$ sec at $T = 10^9$ K. On plugging the known time and cross-section into (9) Gamow found that at $T = 10^9$ K the matter density should be $n \sim 10^{18}$ nucleons cm^{-3}. By equations (4) and (I-14) this translates to $T_0 \sim 10$ K at density 10^{-6} cm^{-3}, which is roughly what the present density is thought to be.

Gamow's calculation was repeated in more detail by Alpher and Herman.[6] For the first time, they explicitly estimated what the radiation temperature would be now. Their result was $T_0 \sim 5$ K. In 1964 Smirnov rediscussed this parameter. He argued that if the element production in the Big Bang were to be anywhere near reasonable the mass density in matter at the epoch where the radiation density is 1 g cm^{-3} should be in the range of 10^{-4} to 10^{-8} g cm^{-3}. This translates to $T_0 = 1$K to 30 K at $\rho_0 = 10^{-30}$ g cm^{-3}.[6] This was compared with possible observational limits on the microwave background temperature by Doroshkevich and Novikov.[6]

The close agreement of the Gamow-Alpher-Herman temperature of the residual radiation from the "Ylem" with the effective temperature $T_0 = 2.7$ K of the modern candidate for the "Primeval Fireball" is an impressive result, but unfortunately it is somewhat delicate for use as an argument in support of the Primeval Fireball hypothesis. In the most direct "naive Big Bang" model the element building process still is thought to operate as far as helium, although it effectively stops there due to the gap at mass 5, as Alpher had suggested from the start. The predicted primeval helium abundance is in the range of 25-30 percent by mass, depending on parameters. It is still by no means clear whether or not the Galaxy started out with this much helium. If it did not there are many ways by which the predicted initial helium may be eliminated without abandoning the 2.7 K Fireball (Chapter VIII). What if the initial helium were in fact high, 25-30 percent, and we had a reliable measure of this number? One might then try to use the naive Big Bang model and the knowledge of this one parameter, the helium abundance, to determine one other parameter, like the present radiation temperature T_0. However, one would find a very large uncertainty in the predicted T_0 because, contrary to what was assumed in equation (9), the neutron-proton abundance ratio is now estimated to be only ~ 0.2 at $T = 10^9$ K. If $T_0 = 2.7$ K most of these neutrons are consumed, producing $\sim$ 25-30 percent helium by mass. If T_0 were 1 K instead of 2.7 K the density at 10^9 K would be *increased* by a factor $(2.7)^3$, because we have to trace the expansion further back to reach 10^9 K, but because most of the neutrons were consumed anyway the primeval helium would be only slightly increased. On the other hand it is amusing to note that if T_0 had been greater than 30 K the nucleon density at $T \sim 10^9$ K would have been *lower* by a factor $> (30/2.7)^3$, and this would have appreciably reduced the neutron capture probability, reducing the primeval helium, and leaving an uncomfortably large residual deuterium abundance. To this sort of accuracy it might be claimed that the fireball temperature is predicted *if* we can establish a reliable measure of the initial helium abundance *and if* this abundance is high, 25-30 percent by mass, comparable to the (rather uncertain) Solar helium abundance. Further details are given in Chapter VIII.

The microwave background first was measured as an anomalous excess instrumental noise, amounting to about 3 K, in the radio telescope used in the first Telstar communication experiments. Penzias and Wilson studied this telescope with care, and could find no local source for the anomaly. That is, it appeared that the radiation must be extraterrestrial. At the same time, at the suggestion of R. H. Dicke, Roll and Wilkinson at Princeton were building a radiometer in the hope of detecting a Primeval Fireball. When the two groups became aware of each other it was quickly decided that Penzias and Wilson possibly had discovered the Primeval Fireball.[7]

The first test of the Primeval Fireball hypothesis evidently is to measure the spectrum, to learn whether it is close to a blackbody distribution. The status of the observations is discussed in the next section. The next obvious measurement is the degree of isotropy of the radiation.[8] This provides a direct test of the assumption that the microwave radiation might originate in the Solar System, or in the Galaxy, or in local sources (Sec. c below). It is also a valuable measure of the isotropy of the Universe (Chapter II) and it should fix our peculiar velocity (Sec. d-i below).

b) *Test of the Fireball Hypothesis – Spectrum Measurements*

i) *Measuring the Fireball*

If the microwave background is blackbody radiation at temperature $T_0 = 2.7$ K the brightness i_ν (eq. 4) should be maximum at wavelength

$$\lambda_m = \frac{0.510}{T_0} = 0.19 \text{ cm}. \tag{10}$$

Another useful function is νi_ν, the brightness per logarithmic frequency interval, which measures where the energy is. This function is maximum at

$$\lambda'_m = \frac{0.367}{T_0} = 0.14 \text{ cm}. \tag{11}$$

When $\lambda >> \lambda_m$ the Planck distribution reduces to the Rayleigh-Jeans form

$$i_\nu = 2k\, T_0\, \nu^2/c^2 . \tag{12}$$

The brightness is directly proportional to temperature in this limiting case.

It might be possible to observe a 2.7 K blackbody distribution from $\sim$ 50 cm wavelength to $\sim$ 0.04 cm wavelength. Longward of 50 cm the blackbody radiation would be lost behind the radio emission from the Galaxy and extra-galactic radio sources. Infrared radiation from dust in the Galaxy is expected to obscure a blackbody distribution somewhere around 0.04 cm.

Longward of 3 mm wavelength there are windows through which the radiation from the atmosphere is reasonably small, so one can make absolute flux measurements from the ground with reasonably small corrections for atmospheric emission. It is customary to measure radiation brightness in terms of an equivalent antenna temperature defined by equation (12),

$$T_a \equiv \frac{c^2 i_\nu}{2k\,\nu^2} .$$

The antenna temperature looking at blackbody radiation coincides with T_0 at long wavelength, $\lambda >> \lambda_m$, and of course T_a/T_0 exponentially decreases at $\lambda << \lambda_m$. The brightness of the atmosphere is roughly constant at 2-3 K, in this measure, longward of 2 cm wavelength, so that the correction for the atmosphere in a ground-based measurement is roughly a factor of 2. It seems to be generally agreed that the coincidence of the atmospheric temperature with T_0 is an accident. The atmospheric radiation is determined by tipping the radiometer beam. The amount of atmosphere it looks through varies as the secant of the angular distance from the zenith, and this is proportional to the atmospheric emission because the atmosphere is optically thin. Thus the extra-terrestial component can be found by curve fitting. Shortward of 3 mm wavelength there apparently are no adequately transparent atmospheric windows, so one must observe above the atmospheric water vapour.

A 2.7 K blackbody distribution would be lost at short wavelength behind the infrared emission from interstellar dust. The wavelength at which this happens is not known with any certainty, but we can establish the following order of magnitude. The energy density of visible starlight in the neighborhood of the Solar System is $u \sim 2 \times 10^{-13}$ ergs cm^{-3}. This may be expressed as an effective temperature by the formula

$$u \equiv aT_e^4 , \quad T_e \cong 2.2 \text{ K} .$$

Again there is a coincidence with T_0, which may or may not be an accident (ref. I-43). It means that locally there is about as much energy in starlight as there is in a 2.7 K Fireball. About half the starlight energy is absorbed by interstellar dust before the radiation can leave the Galaxy, and this energy will be re-radiated somewhere in the infrared to form a background with energy density again roughly comparable to the total energy density of the Fireball. The spectrum of this infrared background is not known; one guess is that it may peak up at $\sim$ 100 microns wavelength and intersect a 2.7 K blackbody distribution at $\sim 400\ \mu = 0.04$ cm wavelength. The number is of course highly uncertain, but with this much reradiated starlight energy somewhere in the infrared it seems clear that one should not expect to be able to trace the Fireball spectrum very far down the Wien tail.

It appears then that a 2.7 K blackbody distribution ought to be observable roughly in the range 50 cm to 0.04 cm wavelength. The features to look for are the Rayleigh-Jeans behavior (eq. 12) $i_\nu \propto \nu^2$ longward of the peak at $\lambda_m \cong 0.2$ cm, and the Wien exponential decrease $i_\nu \propto \nu^2 e^{-h\nu/kT}$ shortward of this.

ii) *Observational Results*

The available observations are listed in Table V-1 and plotted in Figures V-1,2. The antenna temperature in the third column of the table is directly proportional to the energy flux (eq. 12). The thermodynamic temperature listed in the next column is related to the energy flux at a given wavelength by the usual Planck formula (eq. 4). If the spectrum is blackbody the thermodynamic temperature defined in this way is independent of wavelength.

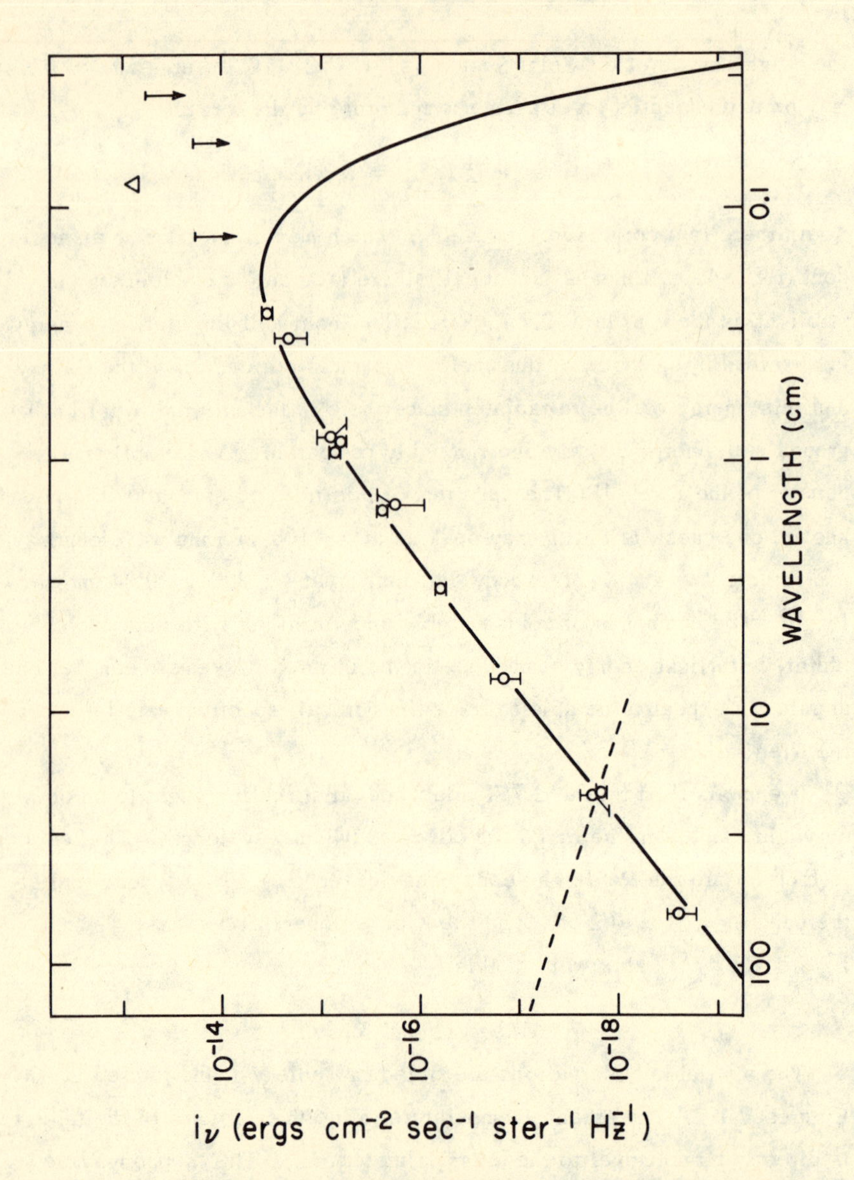

Fig. V-1. The Microwave Background Spectrum.

In the radiometer measurements, lines 1-14 in Table V-1, what is measured is the difference between the antenna temperature (energy flux) looking at the sky and looking into a cold load at liquid helium temperature, $T_{He} \lesssim 4.2$ K (depending on the altitude). As it happens until λ is close to λ_m the difference in energy flux between sky and cold load measured as a difference in antenna temperature is very closely equal to the difference in equivalent thermodynamic temperatures. Wilkinson explains this as follows: the antenna temperature T_a is related to the thermodynamic temperature T by the formula

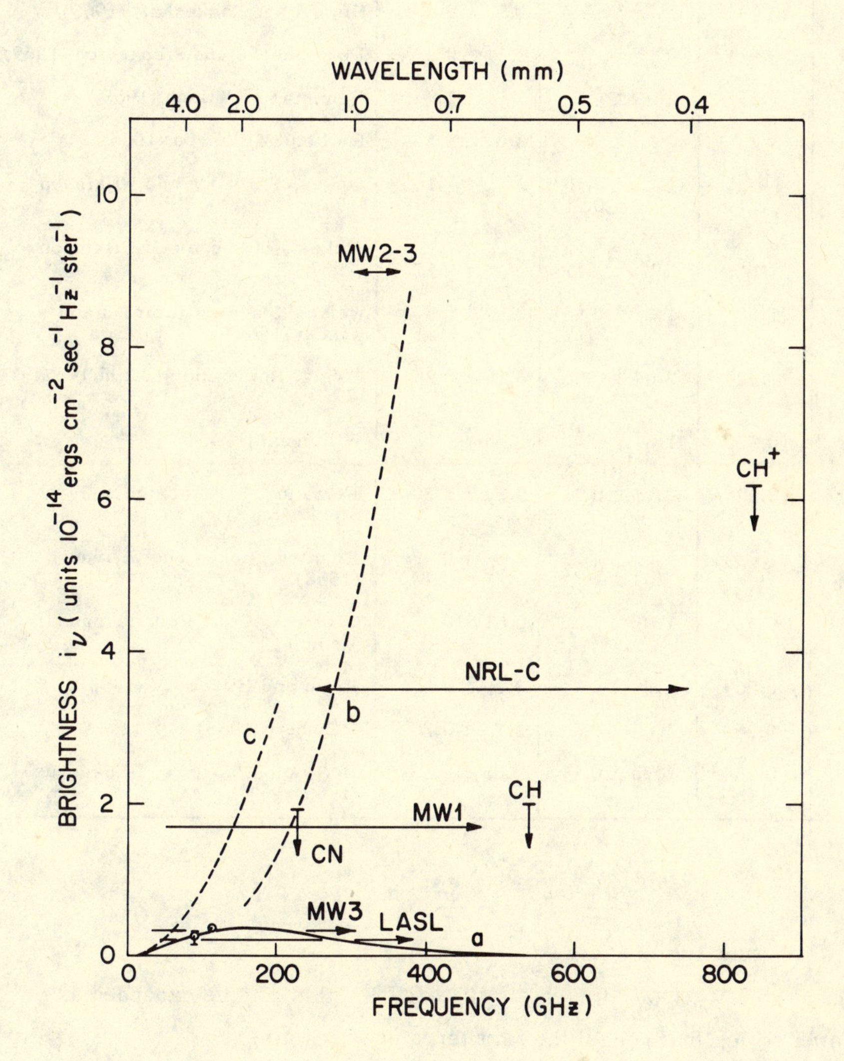

Fig. V-2. The Submillimeter Background Spectrum, after Bortolot, Clauser and Thaddeus.

TABLE V-1
OBSERVATIONS OF THE MICROWAVE BACKGROUND SPECTRUM

	Wave-length (cm)	Antenna Temperature (K)	Thermo-dynamic Temperature (K)	Reference[9]
1	50-75	---	3.7 ± 1.2	Howell and Shakeshaft (1967)
2	21.2	---	3.2 ± 1.0	Penzias and Wilson (1967)
3	20.7	---	2.8 ± 0.6	Howell and Shakeshaft (1966)
4	20.9	---	2.5 ± 0.3	Pelyushenko and Stankevich (1969)
5	7.35	3.0 ± 1.0	3.1 ± 1.0	Penzias and Wilson (1965)
6	3.2	2.8 ± 0.5	3.0 ± 0.5	Roll and Wilkinson (1966)
7	3.2	$2.47^{+0.16}_{-0.21}$	$2.69^{+.16}_{-.21}$	Stokes, Partridge and Wilkinson (1967)
8	1.58	$2.35^{+.12}_{-.17}$	$2.78^{+.12}_{-.17}$	Stokes, Partridge and Wilkinson (1967)
9	1.50	1.56 ± 0.8	2.0 ± 0.8	Welch, Keachie, Thornton and Wrixon (1967)
10	0.924	2.44 ± .26	3.16 ± 0.26	Ewing, Burke and Staelin (1967)
11	0.856	$1.81^{+.16}_{-.21}$	$2.56^{+.17}_{-.22}$	Wilkinson (1967)
12	0.82	2.1 ± 0.7	2.9 ± 0.7	Perzanov, Salmonovich and Stankevich (1967)
13	0.33	0.89 ± 0.32	$2.46^{+0.40}_{-0.44}$	Boynton, Stokes and Wilkinson (1968)
14	0.33	1.01 ± 0.20	2.61 ± 0.25	Millea, McColl, Pederson and Vernon (1971)
15	0.264	1.2 ± 0.4	3.2 ± 0.5	Field and Hitchcock (1966)
16	0.264	1.6 ± 0.6	3.7 ± 0.7	Peimbert (1968)
17	0.264	0.93 ± 0.11	2.83 ± 0.15	Bortolot, Clauser and Thaddeus (1969)

$$\frac{2k\,\nu^2}{c^2}\,T_a = \frac{2h\,\nu^3}{c^2}\,(e^{h\nu/kT}-1)^{-1}\,.$$

When $h\nu/kT < 1$ ($\lambda > 0.5$ cm if $T_0 = 2.7$ K) this may be expanded as a power series in $h\nu/kT$, the first terms of which are

$$T_a = T - \frac{1}{2}\frac{h\nu}{k} + \frac{T}{12}\left(\frac{h\nu}{kT}\right)^2 + \cdots \qquad (13)$$

Because it is a question of difference of antenna temperature looking into sky and cold load, the second term $h\nu/2k$ cancels and the discrepancy between antenna temperature difference and thermodynamic temperature difference shows up only in the third term in the expansion. Even at 0.86 cm wavelength the correction for this third term is only 0.03 K (ref. 9-11).

Line 1 in the table is based on the measured background at 50 and 75 cm wavelength fitted to the form $i_\nu = A\nu^{-0.8} + B\nu^2$. The first term is the expected background from the Galaxy and extragalactic sources, where the exponent 0.8 is determined from the known radio spectrum in the brighter parts of the Galaxy. The second term is an *assumed* component with a Rayleigh-Jeans spectrum. This measurement will be a valuable datum in Chapter VII, but here one should recognize that since there is a large subtraction for the Galaxy, there is a chance for appreciable systematic error. For the measurements near 20 cm wavelength the corresponding correction is only 0.5 K, and still smaller at shorter wavelength.

Lines 13 and 14 are the shortest wavelength ground-based measurements. The greatest correction here is the atmosphere temperature, which amounts to 10-12 K as a linear correction to the antenna temperature at the zenith.

Lines 15-17 are based on the spin temperature of interstellar cyanogen, CN. The observation is the relative number of molecules in the ground and first rotationally excited states as determined from the relative strength of the optical absorption lines resulting from transitions from each of the two states. The history of the discovery and application of this test is given in reference 1.

The observed rotational excitation of CN could be due to some process other than direct radiative transitions at the resonance wavelength, like collisional excitation, or the formation process, or pumping via optical transitions. The main argument against any such process is that the measured excitation temperature is sensibly the same in the directions of at least nine different stars (which must be bright and far away to show interstellar CN absorption). If the excitation effect were a local one it might be expected to vary with the varying local conditions. A second significant point is that the spin temperature does accurately coincide with the temperature deduced from longer wavelength radiometer measurements. As described in part (iv) of this section the weighted mean of the radiometer measurements differs from the CN measure of Bortolot, Clauser and Thaddeus by -0.15 ± 0.16 (s.d.) K, which represents a coincidence to 5 percent accuracy. This is strong evidence for the view that CN and the radiometers are measuring the same thermal phenomenon.

The situation shortward of 1 cm is best illustrated in the linear graph in Figure V-2, which is based on the discussion of Bortolot, Clauser and Thaddeus (ref. 9-17). The 3.3 mm radiometer measurement and the CN spin temperature are represented by the two points in the lower left side of the figure, on a 2.7 K blackbody distribution shown as a solid line. The three points with vertical downpointing arrows are upper limits on the radiation background based on limits found by Bortolot, Clauser and Thaddeus for the excitation temperature of other molecular states. These should be upper limits to the radiation temperature at the wavelengths of the transitions, for any other effect like collisions generally would be thought to increase the spin temperature because the spin system would be interacting with a hotter system.

The direct submillimeter flux measurements are from bolometers above the atmosphere. These measurements cover broad wavelength bands fixed by instrumental response plus filters. The bandwidths of recent observations are roughly indicated by the horizontal lines in Figure V-2, and the height of each line is fixed so that the subtended area is equal to the minimum detected flux. The measurement of the Cornell-NRL groups has been repeated several times.[10] This measurement is from a rocket, as is the Los Alamos result.[10] The measurement by Muehlner and Weiss[11] is from a balloon. It is apparent from the figure that the bolometer results are in strong disagreement with the 2.7 K blackbody distribution. It is by no means clear what this might mean, as is discussed in the next section.

iii) *The Submillimeter Background*

It is premature to draw any conclusions about the significance of the bolometer measurements because, as one must recognize, despite the enormous effort devoted to these measurements they are based on new technology which might yet admit systematic errors. In view of their importance for the Primeval Fireball hypothesis, however, it seems justified to list a few possibilities.

The bolometer measurements may be (a) real, or (b) instrumental bugs. If (a) it appears that the submillimeter flux has a line or band spectrum, as explained below. Also, if (a) the flux (a–1) may be a continuation of the microwave phenomenon, or (a–2) may be a new and independent phenomenon. Possibility (a–1) is the main point of concern here, and it is based on the fact that a power law extrapolation from the microwave measurements gives roughly the flux of the Cornell-NRL group (curve c in Figure V-2).

The band structure of the submillimeter background spectrum is suggested by the fact that Muehlner and Weiss report a strong flux in the band 800 to 1000 μ, indicated by MW 2-3 in the Figure, the energy in the band being roughly one tenth the total energy flux NRL-C, and also by the fact that the NRL-C flux could be consistent with MW 1 and LASL only if there were a strong flux shortward of about 600 μ. Also, the upper bounds from molecular spin temperature measurements could be consistent with the NRL-C result only if the flux were concentrated in bands that happen to miss the two molecular resonances (ref. 9-17).

This last point would not apply if the bolometers were measuring radiation from the ionosphere, say, or from the Solar System. The difficulty with this idea is the reported isotropy of the submillimeter radiation. This applies also to any model for the origin of this radiation within the Galaxy, for we are toward one edge of the Galaxy. Of course, this difficulty would be relieved if one could find some way to scatter the radiation.

The major point of difficulty with interpretation (a–1) is that the "continuity" is not very clear. Unlike the submillimeter background, the microwave spectrum is very smooth, as is apparent from the independent measurements at different wavelengths and from frequency scans in small bands with single instruments. Also, the CN and 3.3 mm radiometer points show that the microwave spectrum does break away from the ν^2 power law spectrum about as it should for a blackbody distribution.

An interesting idea is that the submillimeter background may be the integrated emission from extragalactic sources. Here again there is a problem with isotropy. The optical brightness of the Galaxy is $\sim 30\ S_{10}(V)$ at its thinnest part, looking normal to the disc, and of course it is much greater looking through the Milky Way. By comparison, the integrated light of all the other galaxies is $< 5\ S_{10}(V)$ (Sec. IV-b), so if the Galaxy were a "typical" infrared source it would be expected to dominate the submillimeter background as it does the optical. If the Galaxy is not a typical source the next problem is to make the spectrum of the integrated radiation narrow enough to avoid conflict with the molecular spin temperature limits. For example, let us assume that all galaxies radiate a sharp line at frequency ν_1. Then the integrated intensity from the line satisfies equation (IV-11),

$$\frac{d}{dt}[a(t)^3\, i(t, \nu(t))] = ca(t)^3 \frac{L}{4\pi} \delta(\nu(t) - \nu_1), \tag{14}$$

where L is the mean energy emission rate in the line per unit volume. For the Steady State model the solution to this equation is[12]

$$i_\nu = \frac{cL}{4\pi H\nu_1}\left(\frac{\nu}{\nu_1}\right)^3, \ \nu < \nu_1 . \tag{15}$$

Line b in Figure V-2 is the curve $i_\nu \propto \nu^3$ that just hits the upper bound at the second excited state of CN. Apparently this curve might also be consistent with the suggested flux in the 800-1000 μ band of Muehlner and Weiss if the cut-off ν_1 is about 400 GHz. The limits at CH and CH^+ prevent us from constructing a spectrum of this form that might be consistent with the NRL-C flux.

iv) *Analysis of the Microwave Measurements*

There are 17 measurements of the microwave background listed in Table V-1, enough to bear some statistical analysis. Entries 2-12 in the table will be considered first, as one block of modestly homogeneous data. Entries 15-17 are indirect measures, while entries 1, 13 and 14 required foreground corrections appreciably larger than was the case for the other radiometer data.

An objective measure of how well these data satisfy the expected Rayleigh-Jeans behavior is to fit them to a power law,

$$T = A\lambda^{\beta}. \tag{16}$$

For each measurement λ is known with negligible error, and we have an estimate of the standard deviation δT in the measurement of T, which we can translate to a standard deviation $\sigma = \delta T/T$ on $y = \ln T$. Then the weighted least squares fit of $\ln T = \ln A + \beta \ln \lambda$ gives

$$\beta = \frac{\sum \frac{x_i}{\sigma_i^2} \sum \frac{y_i}{\sigma_i^2} - \sum \frac{x_i y_i}{\sigma_i^2} \sum \frac{1}{\sigma_i^2}}{\left(\sum \frac{x_i}{\sigma_i^2}\right)^2 - \sum \frac{x_i^2}{\sigma_i^2} \sum \frac{1}{\sigma_i^2}},$$

with uncertainty

$$(\delta_\beta)^2 = \frac{\sum \frac{1}{\sigma_i^2}}{\sum \frac{x_i^2}{\sigma_i^2} \sum \frac{1}{\sigma_i^2} - \left(\sum \frac{x_i}{\sigma_i^2}\right)^2},$$

where

$$x = \ln \lambda, \; y = \ln T,$$

and where $\delta\beta$ is the expected standard deviation in β if the σ_i are standard deviations in the measurements y_i.

The result of fitting the thermodynamic temperatures 2 to 12 in Table V-1 to equation (16) is

$$T = 2.80\,\lambda^{(-0.019 \pm 0.034)}\ \mathrm{K}. \tag{17}$$

The result of fitting the antenna temperatures 2-12 to (16) is

$$T_a = 2.22\,\lambda^{(0.073 \pm 0.037)}\ \mathrm{K}. \tag{18}$$

In both equations λ is measured in centimeters. Because antenna temperature is a linear measure of radiation power this last equation says that the radiation brightness fitted to the power law

$$i_\nu \propto \nu^a \tag{19}$$

gives

$$a = 1.937 \pm 0.037,\ 21\ \mathrm{cm} > \lambda > 0.8\ \mathrm{cm}, \tag{20}$$

which deviates from the Rayleigh-Jeans law $i_\nu \propto \nu^2$ by two standard deviations, or four percent; and of course when this is duly corrected for the *expected* deviation from the Rayleigh-Jeans law for a blackbody distribution, as in equation (17), there is no discrepancy to two percent accuracy.

A better way to test for the expected departure from the Rayleigh-Jeans law is to fit the antenna temperatures to the expression

$$T_a = T_0 - \frac{\gamma}{2}\frac{hc}{k\lambda}. \tag{21}$$

By equation (13), $\gamma = 1$ if the distribution is blackbody. The result of fitting the antenna temperatures 2-12 to this expression, weighted according to the indicated uncertainties in the measurements, is

$$\begin{aligned} T_0 &= 2.69\ \mathrm{K}, \\ \gamma &= 0.81 \pm 0.30, \end{aligned} \tag{22}$$

where again ± 0.30 is the standard deviation in γ on the assumption that the indicated errors in the Table are standard deviations.

The result (22) is consistent with the blackbody hypothesis $\gamma = 1$, and it is 2.7 standard deviations away from a straight power law $\alpha = 2$, $\gamma = 0$ (eq. 19). Of course the discrepancy is relieved if we are willing to adjust α.

Going further down the table, we see that the 3.3 mm radiometer points (13) and (14) establish the deviation from the straight power law $\alpha = 2$ to 8 standard deviations. The CN point establishes it to 16 standard deviations. In each case the number of standard deviations of course must be tempered by the possibility of systematic error, but the case for deviation from the power law $\alpha = 2$ by the amount expected on the Fireball hypothesis nonetheless is strong.

The weighted mean value of all the radiometer thermodynamic temperature measurements 1-14 in the table is

$$T_0 = 2.72 \pm 0.08 \text{ K}, \tag{23}$$

and as usual the indicated standard deviation ± 0.08 is based on the assumption that the indicated errors in the individual measurements are standard deviations. An interesting test of this point (and of the model) is to evaluate

$$\chi^2 = \sum (T_i - T_0)^2/(\delta T_i)^2 , \tag{24}$$

where T_i is a measured thermodynamic temperature and δT_i the stated uncertainty in the measurement. The result for the sum over the $N = 14$ radiometer points is

$$\chi^2 = 7.1 . \tag{25}$$

The expected result is

$$\chi^2 = N - 1 \pm (2N - 2)^{1/2} = 13 \pm 5 . \tag{26}$$

The numbers are consistent within the expected uncertainty, so there is no evidence of trouble.

The weighted mean (23) of the radiometer measurements agrees with the CN point of Bortolot, *et. al.*, 2.83 ± 0.15 K, within one standard deviation. The weighted mean of the three CN measures (15) to (17) is 2.89 ± 0.14 K, again consistent with (23) to 5 percent accuracy. This precise coincidence is strong evidence for the Fireball hypothesis.

The weighted mean of all 17 thermodynamic temperature measurements in Table V-1 is

$$T_0 = 2.76 \pm 0.07 \text{ K}. \tag{27}$$

The sum (24) for all measurements is

$$\chi^2 = 10.1\,, \tag{28}$$

and the expected value is

$$\chi^2 = 16 \pm 5\,. \tag{29}$$

The values (25) and (28) for the data possibly are smaller than they ought to be because observers must estimate and add in possible systematic errors in estimating δT, and perhaps also because there is some bias against reporting a number more than one standard deviation from the "right answer." Also, one must remember that stated errors mean different things to different observers. However, if there is a bias it seems not to be large, and the value of χ^2 certainly is consistent with the blackbody hypothesis.

c) *Local Source Model*

A competitor of the Primeval Fireball hypothesis is the local source model,[13] which starts from the very reasonable remark that there are many known extragalactic radio sources, so conceivably the microwave background is the integrated emission from all these sources. Because the microwave background seems to peak around 1 mm wavelength the responsible radio sources should peak there too, and there are examples of such objects. The first question is whether one can give a reasonable account of the spectrum. This has been considered in greatest detail by Wolfe and Burbidge.[13] Following their discussion we can get an impression of what is required by supposing that the sources radiate in a fairly narrow frequency range so that the spectrum of the sources can be considered in first approximation as a delta function. Then equation (14) applies and we find in the Steady State picture $i_\nu \propto \nu^3$, which certainly is not consistent with (19) and (20). In an evolving model L in equation (14) is equal to $n(t)\,\mathcal{L}(t)$, where the source number density $n(t)$ varies as $a(t)^{-3}$ due to expansion, and in addition $n(t)$ and the luminosity per source $\mathcal{L}(t)$ may vary with time as sources form and evolve. Wolfe and Burbidge take this into account by writing

$$L = n(t)\,\mathcal{L}(t) \propto a(t)^{3-m}, \tag{30}$$

where m is a parameter. Then in the Einstein-deSitter cosmology, where $a(t) \propto t^{2/3}$, equation (14) gives

$$i_\nu \propto \nu^{(1.5\,+\,m)}, \tag{31}$$

and equations (19) and (20) require

$$m = 0.437 \pm 0.034\,.$$

If the source spectrum peaks up at 1 mm wavelength the radiation observed now in the range of 0.2-20 cm wavelength would have originated at redshifts $Z = 1$ to $Z = 200$, and during this considerable span of cosmic evolution we must be prepared to assume that the course of evolution of radio sources followed the law (30) (suitably modified to take account of the observed turn-over at the high frequency end) with surprising fidelity.

The second question is the expected irregularity in the background radiation from localized sources. The two interesting cases are (1) that the radiation originated in the "nearby" sources in the Universe about as it is now, and (2) that the radiation originated in sources at high redshift.

In the first case the best calculation of the expected irregularity is given by Gold and Pacini,[13] as follows. A convenient and reasonable model in this case is that the radiation originates in point sources distributed about us at rest in flat space within a sphere of radius $r = cH^{-1}$.

The number density of sources is n_s and the luminosity per source is $\mathcal{L}$. It will be assumed that the sources are distributed at random. The observed quantity is the energy flux F incident within a solid angle Ω. To compute F and its variance suppose the radial distance along the line of sight is divided into narrow intervals, the ith interval being from r_i to $r_i + \Delta r_i$. Then the flux F along a chosen line of sight is

$$F = \sum \frac{n_i \mathcal{L}}{4\pi r_i^2}, \tag{32}$$

where n_i is the number of sources that happen to be within the solid angle Ω and the ith radial interval. The ensemble average value of n_i is

$$\langle n_i \rangle = n_s \Omega r_i^2 \Delta r_i . \tag{33}$$

Because the sum and the ensemble average can be taken in either order the average flux reduces to the familiar expression,

$$\begin{aligned} \langle F \rangle &= \sum \frac{\mathcal{L}}{4\pi r_i^2} \langle n_i \rangle \\ &= \frac{\mathcal{L}}{4\pi} n_s \Omega\, cH^{-1} , \end{aligned} \tag{34}$$

where the maximum radius in the model is cH^{-1} (cf. eq. IV-8). In computing the variance of F the standard trick is to observe that the cross terms in $\langle (F - \langle F \rangle)^2 \rangle$ for different radial intervals are not correlated so they vanish in the ensemble average:

$$\begin{aligned} (\delta F)^2 &\equiv \langle (F - \langle F \rangle)^2 \rangle \\ &= \sum \frac{\mathcal{L}^2}{(4\pi)^2 r_i^4} \langle (n_i - \langle n_i \rangle)^2 \rangle \\ &= \sum \frac{\mathcal{L}^2 \langle n_i \rangle}{(4\pi)^2 r_i^4} \\ &= \frac{\mathcal{L}^2 n_s \Omega}{(4\pi)^2} \int \frac{dr}{r^2} . \end{aligned} \tag{35}$$

The third equation follows from the assumption that the sources are distributed at random, so the variance of n_i is $<n_i>$. The last step then follows from (33).

The integral in (35) diverges as $r \to 0$. The closest sources contribute most of the variance because they are the brightest. There will be an effective cut-off at R_0, say, because the few brightest sources would be recognized as such and eliminated in figuring the variance in the background. With the cut-off at R_0 equations (34) and (35) give

$$\left(\frac{\delta F}{F}\right)^2 \cong \frac{H^2}{n_s \Omega\, c^2 R_0}. \tag{36}$$

Conklin and Bracewell[14] looked for possible irregularity in the sky on an angular scale of 10 arc min, or solid angle

$$\Omega \sim \pi(5')^2 = 7 \times 10^{-6} \text{ ster.}$$

They scanned a strip of the sky about 10^h wide in right ascension, at declination $+41^\circ$, so the total solid angle in their survey was

$$\Omega_T \sim 8 \times 10^{-3} \text{ steradians.}$$

For purposes of discussion let us suppose the 10 nearest sources were noted and eliminated. Then the cut-off R_0 satisfies

$$10 = n_s R_0^3 \Omega_T/3 .$$

The upper limit found by Conklin and Bracewell for the fractional irregularity in the background is

$$\frac{\delta F}{F} < 2 \times 10^{-3}$$

on the angular scale of 10'. Collecting, one finds the minimum source number density

$$n_s \gtrsim 3000\, h^3\, \text{Mpc}^{-3}, \tag{37}$$

and the mean distance between sources

$$n_s^{-1/3} \lesssim 70\, h^{-1}\, \text{kpc}. \tag{38}$$

where h is defined by equation (II-3).

The minimum source number density (37) is some 5 orders of magnitude larger than the number density of large galaxies (eq. IV-3), and the mean distance (38) is about one tenth of the distance to the Andromeda Nebula. It might be noted that the sources are assumed to be randomly distributed, so the source density must be smooth when observed on a scale > 70 kpc, in strong contrast to the galaxies, which are observed to be distributed in a decidedly correlated fashion out to 10 Mpc. That is, the sources must entirely avoid the clustering tendency of galaxies.

A less extreme situation is the second case, where it is assumed that the radiation originated in localized sources at high redshift. We can find a rough estimate of the required source density as follows. Let us adopt the Einstein-deSitter model, and let us express the source density at epoch t as

$$n(t) = n_s \, (a_0/a(t))^3 \, .$$

Then it is readily seen by the methods described in the next Chapter that when $z >> 1$ the number of sources observed in solid angle Ω and at redshift $> z$ is

$$N(> z) = (2c \, H^{-1})^3 \, \Omega \, n_s \, z^{-1/2}, \; z >> 1 \, .$$

If the observed energy flux F came from randomly distributed sources at redshift $\geq z$ the root mean square deviation in F would be roughly

$$\frac{\delta F}{F} \sim N(> z)^{-1/2} \, .$$

Then the Conklin-Bracewell limit implies

$$n_s \gtrsim 0.16 \, h^3 \, z^{1/2} \, \mathrm{Mpc}^{-3} \, .$$

This is larger than the density of large galaxies (eq. IV-3) by a factor of about $10 \, z^{1/2}$.

If the radiation comes from sources at redshift $z > 7$ it may be scattered by intergalactic electrons. This would smooth the radiation and still further reduce the required spatial source density.

It is apparent that the isotropy of the microwave background places a strong but not unavoidable constraint on possible local source models. If the radiation originates in sources around us in the present epoch, as is required in the simplest Steady State picture, then the sources must be much more numerous than "normal" galaxies, and, what seems much more unlikely, the mean distribution of sources must be much more uniform than is the observed large scale (to 10 Mpc at least) distribution of matter in galaxies. If the radiation originated at moderately high redshift, $z > 10$, say, the constraint would be much less strong, and it could be argued also that we need not necessarily look for the highly correlated distribution of matter observed in the present Universe.

d) *Applications of the Primeval Fireball*

The following have been the main proposed applications of the microwave background and the Primeval Fireball hypothesis.

i) *The Aether Drift Experiment*

The microwave radiation provides a frame of reference, for it can appear isotropic only for one preferred motion – an observer moving relative to this preferred frame would find that the radiation is brighter than the average looking in the direction of his motion and dimmer than the average looking in the opposite direction. Given that we are in a sea of blackbody radiation at temperature T_0, it is a straightforward exercise in Lorentz transformations to show that an observer, moving at speed v relative to the preferred frame of the radiation and looking at angle θ from this velocity vector, sees a blackbody spectrum at temperature[15]

$$\begin{aligned} T(\theta) &= T_0\,(1 - v^2/c^2)^{1/2}\,(1 - \frac{v}{c}\cos\theta)^{-1} \\ &\cong T_0\,(1 + \frac{v}{c}\cos\theta),\ v/c << 1. \end{aligned} \tag{39}$$

By fitting the measured temperature as a function of direction to (39) one can determine the direction and magnitude of our peculiar velocity.

One should bear in mind that this experiment does not violate relativity, because we are only determining our velocity relative to the radiation. A similar special frame of motion is provided by the distant galaxies, for only with one preferred motion does the cosmological redshift appear isotropic. This latter effect has been under discussion for some time. There have been attempts to fix our peculiar velocity from it, and of course it will be of considerable interest to discover whether the two measures of peculiar velocity are consistent. It might be noted that this experiment does not depend on the assumption that the microwave background is the Primeval Fireball, only that the source is uniformly distributed. One would look for the same effect in the isotropic X-ray background (although the amplitude $\delta i/i$ would be different, depending on the spectrum).

For some years de Vaucouleurs has been concerned about the possible peculiar velocity of the Local Group inferred from the distribution and radial velocities of galaxies around us.[16] Sciama[17] first discussed this question in connection with the microwave isotropy experiment. It is argued that there can be three important contributions to our peculiar velocity, the velocity of our rotation in the Galaxy, the motion of the Galaxy relative to the center of mass of the Local Group, and the motion of the Local Group relative to the Virgo cluster of galaxies.

Convenient coordinates in this discussion are right ascension and declination, which are the polar angles for a right handed coordinate system with orientation fixed relative to the fixed stars, the polar axis along the North Pole of the Earth. The declination δ is measured from the Celestial Equator, like latitude, so that the standard polar angle is $\theta = 90^\circ - \delta$. The right ascension α is the usual azimuthal angle ϕ measured from a fixed direction in the sky (relative to the fixed stars), with α increasing in the counterclockwise direction looking down from the North Pole, and usually α is expressed in units of hours instead of degrees.

In the original Princeton isotropy experiment[8] the rotation of the Earth was used to scan the radiation brightness near the celestial equator (that is, the scan is in right ascension at fixed declination $\delta = -8^\circ$), the reference being provided by the radiation brightness looking in a fixed direction in the sky, toward the North Pole. Because this is a comparison experiment it can be done much more precisely than the absolute flux measurement. The comparison cancels the atmospheric emission to a first approximation because one is looking through about the same amount of atmosphere in the two directions. Also, the expected effect has a period of 24 hours sidereal time (one rotation of the Earth relative to the fixed stars) while any systematic diurnal contamination would be concentrated at 24 hours Solar time period. Once the data stretch over a year or more the Fourier transform resolves Solar and sidereal periods, and reveals any serious Solar diurnal contamination. The Princeton experiment did not reveal a statistically significant 24-hour sidereal amplitude. The implied limit on the component of our peculiar velocity projected in the celestial equator was about 250 km sec^{-1} (standard deviation). A slightly modified experiment at Yuma, Arizona, with a scan at $\delta = +42^\circ$ as well as $\delta \sim 0$, gave comparable results. Conklin[18] measured the anisotropy over a shorter time but with much lower instrumental and atmospheric noise, and obtained a tentative positive result, with 24-hour amplitude of the background temperature of $(1.9 \pm 0.8) \times 10^{-3}$ K, maximum at 10 h R.A., for a scan at declination $\delta = +32^\circ$. This projects to velocity components

$$\begin{aligned} v_x &= -220 \pm 75 \text{ km sec}^{-1}, \\ v_y &= +124 \pm 75 \text{ km sec}^{-1}, \end{aligned} \tag{40}$$

where as usual the positive x axis is along $\alpha = \delta = 0$, the positive y axis along $\alpha = 90^\circ$, $\delta = 0$. The major uncertainty in this measurement appears to be the correction for radio emission by the Galaxy, which is a factor ~ 3 larger than (40), and was obtained by scaling from radio intensity maps at longer wavelength according to the law $T_G \propto \lambda^{2.8}$.

In another experiment Henry[19] used a balloon to carry a radiometer above atmospheric noise, and attempted to scan the whole visible sky. His equatorial component is consistent with Conklin's result although his resolution in this direction is not good, and he presents evidence that the z component of the velocity is

$$v_z \cong -200 \pm 100 \text{ km sec}^{-1} . \qquad (41)$$

Henry's measurement was at 3.0 cm wavelength, Conklin's at 3.75 cm, so the correction for the Galaxy in (41) is smaller than for (40) by a factor $(3.75/3.0)^{2.8} \cong 2$.

The components (40) and (41) must be considered preliminary and tentative results, and as such do not bear detailed analysis. It is interesting, however, to write down for comparison some known and some speculated contributions to our peculiar velocity. The first is our (approximately) circular motion in the disc of the Galaxy, rotation speed $\Theta \cong 250$ km sec^{-1}. The Cartesian components of this velocity are shown in Table V-2. The next velocity in the table is based on the assumption that we are falling toward M31, the Andromeda Nebula (cf. Sec. IV-c-ii). M31 apparently is appreciably more massive than the Galaxy, so we have taken the falling velocity as the observed relative radial velocity (corrected for rotation of the Galaxy). The last velocity in the table is directed toward the Virgo cluster, the thought here being that the mass concentration in that direction may have slowed our motion away from it. The magnitude of this velocity has been written as $1000\,\epsilon$ km sec^{-1}, so that ϵ may be interpreted as the fractional deviation from Hubble's law for our recession from the cluster.

TABLE V-2

POSSIBLE CONTRIBUTIONS TO OUR PECULIAR VELOCITY*

	v_x	v_y	v_z
Rotation ($\Theta = 250$)	123	−113	186
To M31, v = 90	67	12	59
To Virgo, v = 1000ϵ	-970ϵ	-120ϵ	220ϵ

*unit = km sec^{-1}

It is apparent that the one free parameter ϵ is not enough to allow us to adjust the sum of the velocities in Table V-2 to (40) and (41), and in particular v_z from the table has the wrong sign according to (41) if we want to keep ϵ positive as is suggested by the above interpretation of this term.

The experimental situation here still is quite unclear, the results (40) and (41) highly preliminary, because the statistical accuracy is poor and the Galaxy correction large and uncertain. The theoretical situation also is unclear because we do not understand the Local Group (IV-c-iii) or the dynamics of the Local Supercluster.

ii) *Interaction with High Energy Cosmic Rays*

Very energetic cosmic ray protons are expected to suffer an appreciable drag force in the Fireball radiation because a fireball photon in the proton rest frame can exceed the threshold for photo-pion production.[20] We can understand the order of magnitude of the effect as follows. The center of mass energy of the photon and proton is

$$q' + m_p c^2 = (m_p{}^2 c^4 + 2Eq - 2\,pqc \cos\theta)^{1/2}$$

$$\cong \frac{Eq}{m_p c^2}(1 - \cos\theta) + m_p c^2 ,$$

where E is the proton energy, q the photon energy, and θ the angle between the momenta, all in the comoving coordinate frame. On taking the typical energy of a Fireball photon to be $q \sim 3kT_0$, $T_0 = 2.7$ K, we find that q' exceeds the π production threshold 135 MeV when

$$E \gtrsim 1 \times 10^{20}\ \text{eV},$$

and reaches the strong peak in the production cross-section at 300 MeV when

$$E \sim 3 \times 10^{20}\ \text{eV}.$$

This is just comparable to the highest energies observed in extensive air shower events, which are presumed to be single protons.

Next we have to consider the rate of loss of energy of the proton. The blackbody photon density is given by equation (7). The cross-section for pion production is $\sigma \sim 10^{-28}$ cm^2. The pion production in the proton rest frame is roughly isotropic, so the mean proton energy loss in the comoving frame is $\sim q'E/m_p c^2$. The product of these factors gives the proton energy loss rate,

$$-\frac{d}{dx}\ln E \sim \sigma n_\gamma q'/m_p c^2$$
$$\sim (30\ \mathrm{Mpc})^{-1}, \tag{42}$$

at $E \sim 3 \times 10^{20}$ eV. The characteristic decay length defined by equation (42) is about 3 times the distance to the Virgo cluster of galaxies. At higher energy E this length is somewhat increased because q' moves past the peak in the production cross-section.

For a more accurate number one should integrate the blackbody distribution against the cross-section and energy loss factor. This has been done by Stecker,[21] who finds

$$\left[-\frac{d}{dx}\ln E\right]^{-1} = 200\ \mathrm{Mpc\ at}\ E = 1 \times 10^{20}\ \mathrm{eV},$$
$$= 40\ \mathrm{Mpc\ at}\ E = 2 \times 10^{20}\ \mathrm{eV}, \tag{43}$$
$$= 15\ \mathrm{Mpc\ at}\ E = 1 \times 10^{21}\ \mathrm{eV}.$$

The last line represents the minimum in the attenuation length. Again, the characteristic distance is large compared to the distance between galaxies, small compared to cH^{-1}.

The next question evidently is, where do the cosmic ray particles come from? We are guided here by the fact that the radius of curvature of the path of a 10^{20} eV proton in a field 10^{-6} Gauss, roughly the interstellar magnetic field, is

$$R = \frac{E}{Be} \sim 100\ \mathrm{kpc},$$

about 10 times the radius of the Galaxy. That is, energetic protons would travel in very nearly straight lines through the Galaxy, so if they originated in the Galaxy they would be expected to come predominantly from one hemisphere, because we are on one side of the Galaxy, and they might even be expected to come from the Milky Way, because that is where the young and active stars (and pulsars) are concentrated. The very energetic cosmic ray particles show no such correlation in incident direction, so we conclude that they must be extragalactic.

If the energetic cosmic rays have been accumulating over the characteristic time H^{-1} then on the Fireball picture we would have expected to find a strong shoulder in the energy distribution function setting in at $\sim 5 \times 10^{19}$ eV, because less energetic particles are accumulated for the whole Hubble distance (or equivalent time) $cH^{-1} \sim 3000$ Mpc, while particles at 2×10^{20} eV could have accumulated from only those sources within ~ 40 Mpc. This appears not to be the case – the tentative indication is that the observed spectrum of extensive air shower event energies is smooth and continuous through this region.[22] We do not understand this.

It does seem clear from the isotropy measurements that the microwave background is extragalactic and universal. The background spectrum is well established longward of about 3 mm wavelength, and this measured part of the spectrum includes 35 percent of the photons in a 2.7 K blackbody spectrum. Thus if it were assumed that the microwave background fell below the blackbody spectrum shortward of 3 mm it would increase the proton mean free path, but not enough to avoid the problem. Going the other way, if the energy flux at ~ 1 mm wavelength reported by Muehlner and Weiss were extragalactic it would make the photon number density a factor of 3 greater than blackbody, and reduce the distances (43) by the same factor, bringing the minimum mean free path close to the mean distance between giant galaxies. That is, the problem is with us however we interpret the submillimeter background. It is conceivable that these particles were produced in some ''local'' explosion less than 10^8 y ago, and have been trapped in a region of less than about 10 Mpc or so across by an intergalactic magnetic field. It is conceivable also that we have identified extensive air shower events with the wrong particle, although we have no ready alternative candidate.

iii) *Miscellaneous*

A number of other possible applications have been studied in more or less detail. The following are some examples.

1) The Primeval Fireball would unambiguously rule out the simple-minded Steady State model because there is no place for a dense epoch when the radiation could have been thermalized. A variant mentioned in Chapter I (ref. I-44) is that the creation happens in bursts, between which the Universe expands and evolves much like the Friedman model. If the burst ends up at high enough energy it could be accompanied by radiation that, depending on conditions, may be close to blackbody.

2) In the Lemaître model the entropy per nucleon (eq. 6) is very nearly constant, in principle can be measured, and from it we should be able to deduce some elements of the thermal history of the Universe. The main observable application that has been studied so far is the primeval helium production (Chapter VIII). If and when the Fireball spectrum can be measured with sufficient accuracy we should be able to use this information as a probe of departures from the simple homogeneous isotropic and equilibrium picture assumed so far. Some preliminary discussion of this question is given in Chapter VII.

3) The interaction of cosmic ray particles with the Fireball radiation has been mentioned above. Another process that has received attention is "inverse Compton scattering," whereby an energetic electron scatters a Fireball photon, producing an X-ray photon. Hoyle, Gould, and Felten[23] have argued that this process might produce the isotropic X-ray background (Sec. IV-e).

4) The possible discovery of the Primeval Fireball has stimulated a good deal of theoretical study of galaxy formation in an expanding Universe. This is because the radiation plays an important role in the evolution of density irregularities, so one can make some progress without having to come to grips with the vexatious problem of initial conditions.[24]

e) *Is This the Primeval Fireball?*

The Primeval Fireball hypothesis is about as well established as any of the conventional tenets of cosmology, including the hypothesis of the expansion of the Universe. This is not to say that either hypothesis is known to be valid beyond reasonable doubt, or even to deny that there are clouds on the horizon in either case, only that both have appreciable observational support and are reasonable and fertile bases for further discussion. The expansion hypothesis has been under discussion for a much longer period of time, which is significant but hard to quantify. We do have the following objective points, which may be compared to the list in Section I-e.

1) The microwave background spectrum i_ν is proportional to ν^α, where the index α fits, with no adjustable parameters, the blackbody distribution, to an accuracy of 2 percent.

2) There is evidence that the spectrum breaks away from the power law at short wavelength as expected for a blackbody distribution. For $\lambda \geq 0.8$ cm the departure for the power law is specified by the parameter $\gamma = 0.81 \pm .30$, assuming $\alpha = 2$, consistent with the blackbody distribution $\gamma = 1$, and 2.7 standard deviations from the pure power law case $\alpha = 2, \gamma = 0$. The 3.3 mm radiometer measurements clearly show this break, and so does the indirect but apparently very precise CN value.

3) The phenomenon is established over a broad range in wavelength, from ~ 20 cm to 0.25 cm.

4) The radiation is isotropic to good precision, as expected in the simplest model (although it can be argued that it is surprising that the simple model should be so accurate).

There are two clouds on the horizon:

1) The submillimeter background may not fit the hypothesis. It is unclear whether this effect is real, or, if real, whether it is related to the microwave background.

2) The spectrum of very high energy cosmic rays is smooth and continuous, which is surprising if the microwave radiation shortward of one cm is extragalactic and the cosmic rays are protons.

There are two major points on which the weight of evidence for the expansion hypothesis clearly is greater than the weight of evidence for the Fireball hypothesis.

1) One argues that the numerical constant H^{-1} derived from Hubble's law is "reasonable and expected" within a factor of perhaps 2 in total range (which is the accuracy with which H seems to be known) in light of other age estimates, and, to much less accuracy, in light of the characteristic number $(G\rho)^{-1/2}$ derived from estimates of the mean mass density. The best we can say for $T_0 = 2.7$ K is that it is in the right range, within a factor of perhaps 10 either way, to make what may or may not be a reasonable amount of primeval helium.

2) There is no serious competitor to the expansion hypothesis. It is true that one talks of a tired light effect, but there is no known physical process to substantiate it. There is an alternative to the Primeval Fireball hypothesis, in the local source model, and there really are extragalactic radio sources. As we have described, the difficulty here is with the details of fitting the model to the observations.

REFERENCES

1. P. J. E. Peebles and D. T. Wilkinson, *Sci. American* 216, June, 1967; P. J. E. Peebles, *J. Roy. Astron. Soc.* (Canada) 63, 4, 1969; R. A. Alpher and R. Herman, *Reflections on Big Bang Cosmologies*, F. Reines, ed. (Cambridge), 1971.
2. R. C. Tolman, *Relativity Thermodynamics and Cosmology*, (Oxford) 1934; § 171.
3. R. A. Alpher, H. A. Bethe and G. Gamow, *Phys. Rev.* 73, 803, 1948; R. A. Alpher and R. C. Herman, *Rev. Modern Phys.* 22, 153, 1950 and earlier references therein.
4. G. Gamow, *J. Wash. Acad. Sci.* 32, 353, 1942; *Phys. Rev.* 70, 572, 1946.
5. G. Gamow, *Phys. Rev.* 74, 505, 1948; *Nature* 162, 680, 1948.
6. R. A. Alpher and R. C. Herman, *Nature* 162, 774, 1948; Yu. N. Smirnov, *Astron. Zh.* 41, 1084, 1964; Engl. tr. *Soviet Astronomy – A.J.* 8, 864, 1965; A. G. Doroshkevich and I. D. Novikov, *Dokl. Akad. Nauk. SSSR* 154, 745, 1964; Engl. Tr. *Soviet Physics – Doklady* 9, 111, 1964.

7. A. A. Penzias and R. W. Wilson, *Ap. J.* 142, 419, 1965; R. H. Dicke, P. J. E. Peebles, P. G. Roll and D. T. Wilkinson, *Ap. J.* 142, 414, 1965.

8. R. B. Partridge and D. T. Wilkinson, *Phys. Rev. Letters* 18, 557, 1967; *Nature* 215, 719, 1967; P. J. E. Peebles, Lectures in Applied Mathematics 8, Relativity and Cosmology, J. Ehlers, ed., 1967, p. 274.

9. The references to Table V-1 are: (1) T. F. Howell and J. R. Shakeshaft, *Nature* 216, 753, 1967; (2) A. A. Penzias and R. W. Wilson, *A. J.* 72, 315, 1967; (3) T. F. Howell and J. R. Shakeshaft, *Nature* 210, 1318, 1966; (4) S. A. Pelyushenko and K. S. Stankevich, *Astron. Zh.* 46, 223, 1969; Engl. tr. *Soviet Astronomy – A. J.* 13, 223, 1969; (5) A. A. Penzias and R. W. Wilson, *Ap. J.* 142, 419, 1965; (6) P. G. Roll and D. T. Wilkinson, *Phys. Rev. Letters* 16, 405, 1966; (7),(8) R. A. Stokes, R. B. Partridge and D. T. Wilkinson, *Phys. Rev. Letters* 19, 1199, 1967; (9) W. J. Welch, S. Keachie, D. D. Thornton and G. Wrixon, *Phys. Rev. Letters* 18, 1068, 1967; (10) M. S. Ewing, B. F. Burke and D. H. Staelin, *Phys. Rev. Letters* 19, 1251, 1967; (11) D. T. Wilkinson, *Phys. Rev. Letters* 19, 1195, 1967; (12) V. I. Puzanov, A. E. Salmonovich and K. S. Stankevich, *Astron. Zh.* 44, 1129, 1967; Engl. tr. *Soviet Astronomy – A. J.* 11, 905, 1968; (13) P. E. Boynton, R. A. Stokes and D. T. Wilkinson, *Phys. Rev. Letters* 21, 462, 1968; (14) M. F. Millea, M. McColl, R. J. Pederson and F. L. Vernon, Jr., *Phys. Rev. Letters* 26, 919, 1971; (15) G. B. Field and J. Hitchcock, *Ap. J.* 146, 1, 1966; (16) M. Peimbert, *Bull. Obs. Tonantzintla* No. 30, 1968; (17) V. J. Bortolot, Jr., J. F. Clauser and P. Thaddeus, *Phys. Rev. Letters* 22, 307, 1969.

10. J. L. Piper, J. R. Houck, B. W. Jones, and M. Harwit, *Nature* 231, 375, 1971, and earlier references therein; A. G. Blair *et. al.*, Los Alamos Scientific Laboratory.

11. D. Muehlner and R. Weiss, *Phys. Rev. Letters* 24, 742, 1970.

12. R. J. Gould and D. W. Sciama, *Ap. J.* 140, 1634, 1964. For a discussion of cosmological interpretations of the submillimeter background see L. J. Caroff and V. Petrosian, *Nature* 231, 378, 1971.

13. D. W. Sciama, *Nature* 211, 277, 1966; D. Layzer, *Astrophysics Letters* 1, 99, 1968; J. V. Narlikar and N. C. Wickramasinghe, *Nature* 217, 1236, 1968; T. Gold and F. Pacini, *Ap. J.* 152, L115, 1968; A. M. Wolfe and G. R. Burbidge, *Ap. J.* 156, 345, 1969.
14. E. K. Conklin and R. N. Bracewell, *Nature* 216, 777, 1967; *Phys. Rev. Letters* 18, 614, 1967.
15. P. J. E. Peebles and D. T. Wilkinson, *Phys. Rev.* 174, 2168, 1968.
16. G. de Vaucouleurs and W. L. Peters, *Nature* 220, 868, 1968, and earlier references therein.
17. D. W. Sciama, *Phys. Rev. Letters* 18, 1065, 1967; J. M. Steward and D. W. Sciama, *Nature* 216, 748, 1967.
18. E. K. Conklin, *Nature* 222, 971, 1969, and unpublished results.
19. P. S. Henry, *Nature* 231, 518, 1971.
20. R. H. Dicke and P. J. E. Peebles, *Space Science Reviews* 4, 419, 1965; K. Greisen, *Phys. Rev. Letters* 16, 748, 1966; G. T. Zatsepin and V. A. Kuz'min, *Zh. E.T.F. Pis'ma* 4, 114, 1966; Engl. tr. *J.E.T.P. Letters* 4, 78, 1966.
21. F. W. Stecker, *Phys. Rev. Letters* 21, 1016, 1968.
22. I am grateful to C. B. A. McCusker for his information on the EAS data and its possible meaning.
23. F. Hoyle, *Phys. Rev. Letters* 15, 131, 1965; R. J. Gould, *Phys. Rev. Letters* 15, 511, 1965; J. E. Felten, *Phys. Rev. Letters* 15, 1003, 1965.
24. eg. P. J. E. Peebles and J. T. Yu, *Ap. J.* 162, 815, 1970, and earlier references therein.

VI. A CHILD'S GARDEN OF COSMOLOGICAL MODELS

A central goal of cosmology has been to test cosmological models by looking for relativistic effects in the appearance of distant galaxies. The first major observational effort was Hubble's counts of galaxies as a function of limiting apparent magnitude. More recently Sandage has considered in detail the relation between redshift and apparent magnitude.[1] In first approximation the energy flux f from a galaxy of fixed luminosity should vary as z^{-2} ($f \propto \ell^{-2}$ by the inverse square law, $\ell \propto z$ by Hubble's law). In the next approximation the deviation from this law determines one parameter which (if general relativity with $\Lambda = 0$ is valid, and if the homogeneous isotropic world picture is valid) fixes the value and sign of the term $R^{-2} a_0^{-2}$ in equation (I-10). If negative or zero it means that the Universe is open, has infinite volume, and is fated to keep on expanding forever, if positive that the Universe is closed, finite, and doomed to collapse back in on itself.

It is important to distinguish two aspects of the tests. We very much need a test of the hypothesis of the expansion of the Universe, and concordant results among the Hubble test, the Sandage test, the mass density, and so on, would provide it. The second aspect is that the tests may determine the sign of R^{-2}. When we attempt to trace the evolution of the Universe back in time, as in the next chapter, we find that the value of R^{-2} enters as an important parameter, but that there is no discontinuity in the history of the Big Bang models when R^{-2} goes from positive to negative, and in fact unless we are willing and able to trace back beyond the singularity at $a = 0$ there is no way we can discover whether the Universe is infinite or finite from the detailed history of the Universe (except of course by actually determining the value of the parameter R^{-2}!). The great difference between the open and closed models is in the future evolution. However, until cosmology is a good deal more mature than it is now the emphasis must be on a tentative reconstruction of what may have happened based on the few bits of fossil evidence, not on what is going to happen.

The purpose of this chapter is to give the theoretical basis of the Lemaître cosmological model, and to summarize the prospects for observational tests of the model. More complete theoretical explanations will be found in the books of Tolman, McVittie, Landau and Lifshitz, and Robertson and Noonan.[2] On the observational side an important reference is Sandage's study of how the 200 inch telescope might best be used to test cosmological models.[3]

In this chapter equations will be simplified by choosing units such that the velocity of light is unity.

a) *Derivation of the Lemaître Model*

i) *Coordinate Systems*

It will be recalled that a solution $g_{ij}(x)$ to Einstein's field equations has a good deal of uninteresting information on how the coordinate labels x^i have been applied to points (events) in space-time. It does have the following meaning. If x^i and $x^i + dx^i$ are the coordinate labels of two neighboring points in space-time then we can form the scalar line element

$$ds^2 = g_{ij}\, dx^i\, dx^j . \tag{1}$$

If $ds^2 > 0$, ds is defined to be the time interval recorded by a real clock as it moves freely from x^i to $x^i + dx^i$. If $ds^2 < 0$, by definition $|ds^2|^{1/2}$ is the length of a piece of string one end of which hit the point x^i and the other end the point $x^i + dx^i$ simultaneously as judged by an observer sitting on the string. By assumption all rods and clocks give consistent measures of ds^2. If $A^i(x)$ and $B^i(x)$ are two vector fields, then

$$g_{ij}(x)\, A^i(x)\, B^i(x)$$

is the ordinary Lorentz scalar product of A and B as determined by an observer at x.

The path of a free point particle moving between given initial and final positions is fixed by the condition that the action

$$I = -m \int_i^f ds$$

be a minimum. The Euler-Lagrange equations for this extremum problem are the geodesic equations of motion,

$$\frac{d}{ds}(g_{ij}\, u^j) = \frac{1}{2} g_{jk,i}\, u^j u^k , \tag{2}$$

where the vector four velocity of the particle is

$$u^i = dx^i/ds . \tag{3}$$

There is enough freedom of coordinate transformation that one can always arrange that, along the path of a freely moving particle, g_{ij} is equal to the diagonal from $(1, -1, -1, -1)$ and all the first derivatives of g_{ij} vanish.[4] This is called a locally Minkowski coordinate system. Its importance is that all the laws of physics that do not depend on space curvature reduce to the ordinary flat space gravity-free form we all know and love. This is the coordinate system a freely moving observer would define in his neighborhood with his measuring rods and clocks – for example, ds^2 (eq. 1) reduces to the usual Lorentz invariant interval $dt^2 - d\ell^2$. A coordinate system with coordinate positions x^α at rest relative to the Earth is not locally Minkowski because we know in this coordinate system free bodies fall (eq. 2), and we know clocks differing in elevation by h have rates discrepant by the fractional amount gh/c^2. It is not always possible to remove the second derivatives of g_{ij} by coordinate transformation – neighboring geodesics can diverge (eq. 2), and that is why we have tides.

We will be considering two convenient choices of coordinates. Comoving coordinates for the Universe were defined in two ways in Sections I-a and V-e-i. In the present context it is imagined that matter may be described as a continuous fluid, and that each fluid element has attached to it three spatial coordinate labels x^{α}. Then a point in space-time is labelled by the coordinates x^{α} belonging to the fluid element passing through the point and by a fourth chosen time parameter specified along the path of each fluid element. These are comoving coordinates.

In a time-orthogonal coordinate system one starts with some given space-like hyper-surface σ (a three-dimensional surface with normal n^i everywhere time-like, $g_{ij}\, n^i\, n^j > 0$) and with a set of freely moving observers with velocities at σ normal to σ. Each of these observers is assigned a fixed spatial coordinate label x^{α}, and each observer carries a clock set to read some standard time t_i at σ. Then any point in space-time will be labelled by the three coordinates x^{α} of the observer that passes through the point and by the observer's clock reading t at the point. Because coordinate time is proper time for these observers the line element taken along the path of one of the observers ($dx^{\alpha} = 0$) is $ds^2 = dt^2$, so

$$g_{00} = 1 . \tag{4}$$

The four velocity of one of the observers in the time-orthogonal coordinates is

$$u^{\alpha} \equiv \frac{dx^{\alpha}}{ds} = 0,\ u^0 \equiv \frac{dx^0}{ds} = \frac{dt}{ds} = 1 . \tag{5}$$

This velocity is normal to σ, which means that, at σ,

$$g_{\alpha 0} = 0 . \tag{6}$$

With equations (4) and (5) the geodesic equations of motion (2) of one of the freely moving observers are simply

$$\frac{d}{dt} g_{\alpha 0} = 0 ,$$

so (6) has to hold at all times. The line element is then

$$ds^2 = dt^2 - g_{\alpha\beta}\, dx^{\alpha}\, dx^{\beta} . \tag{7}$$

If one wants to choose coordinates that are both time-orthogonal and comoving evidently the matter must move along geodesics, so it cannot suffer any non-gravitational force like a pressure force. This can be arranged if the pressure of the material is negligibly small, or if the model is homogeneous, so that the pressure gradient vanishes. Also, it is apparent that if coordinates can be chosen to be comoving and time-orthogonal then the matter velocity field expressed in a general coordinate system must satisfy the covariant equation

$$u_{i;j} - u_{j;i} = u_{i,j} - u_{j,i} = 0\,, \tag{8}$$

because the expression certainly vanishes in the coordinates of equation (5), and, since the expression is covariant, it has to vanish in any other coordinate system. Conversely when (8) is satisfied one can construct the needed hypersurface σ with normal everywhere parallel to u^i. To see this, suppose σ is defined by the equations

$$x^i = x^i(\lambda^\alpha)\,,$$

where the 3 parameters λ^α specify position in the 3-dimensional surface σ. Any displacement in σ has to be perpendicular to the matter velocity,

$$u_i \, \partial x^i / \partial \lambda^\alpha = 0\,. \tag{9}$$

This is a differential equation for the functions $x^i(\lambda^\alpha)$, and the condition for integrability is

$$\frac{\partial^2 x^i}{\partial \lambda^\alpha \, \partial \lambda^\beta} = \frac{\partial^2 x^i}{\partial \lambda^\beta \, \partial \lambda^\alpha}.$$

When (9) is differentiated with respect to λ^β, the integrability condition gives

$$(u_{i,j} - u_{j,i}) \frac{\partial x^i}{\partial \lambda^\alpha} \frac{\partial x^i}{\partial \lambda^\beta} = 0\,,$$

which is satisfied if (8) is valid.

If equation (8) is not satisfied it means that an observer would see the matter about him rotating relative to his inertial frame defined say by two gyroscopes. Such cosmological models with rotation, along with models that are irrotational but have shear motion in the fluid, have attracted attention in studies of the possible state of the early Universe.[5] Our direct evidence on the rotation of the system of galaxies about us is poor. For example, the nearest giant galaxy, the Andromeda Nebula, is thought to be 700 kpc away and moving toward the Galaxy at 100 km sec^{-1}. If the transverse velocity were close to the velocity of light, 3×10^5 km sec^{-1}, its observed proper motion would be only 10 arc sec per century. However, as described in Chapter II, the very precise isotropy of the radiation background suggests that the Universe is spherically symmetric to good accuracy, which would imply that rotation is unimportant.

ii) *The Robertson-Walker Line Element*

The form of the line element may be obtained by symmetry arguments from the assumption of homogeneity and isotropy, as was shown by Robertson and Walker.[6] The homogeneity means that through any point in space-time there is a three-dimensional space-like hypersurface within which the mass density is constant and the observed local expansion rate is everywhere the same. The Universe looks the same everywhere within this slice of space-time. If the Universe is expanding there is supposed to be a sequence of hyper-surfaces of homogeneity, each belonging to a different density, and each surface can be labelled with a cosmic time variable t. The matter velocity at any point has to be perpendicular to the surface of homogeneity through the point, for otherwise one could define a preferred direction in σ, contrary to the isotropy assumption. Each fluid element can be assigned a fixed coordinate label x^α, and since each fluid element must be moving freely we arrive by construction at time-orthogonal comoving coordinates, x^α, t, with density and expansion rate functions of cosmic time t alone.

Next, let us set up a polar coordinate system. The radial coordinate will be r, where a surface of fixed r and t is an ordinary two-dimensional sphere, a surface at constant proper distance from the chosen origin. The polar angles θ and ϕ are the usual angles an observer at the origin would set up, and by the spherical symmetry the proper length subtended on a sphere of fixed r by chosen angle at the origin has to vary as $(d\theta^2 + \sin^2\theta\, d\phi^2)^{1/2}$. Thus the line element looks like

$$ds^2 = dt^2 - A(r,t)\, dr^2 - B(r,t)\,(d\theta^2 + \sin^2\theta\, d\phi^2). \tag{10}$$

Again the isotropy means that $g_{r\theta}$ and $g_{r\phi}$ vanish, and that A and B can be functions of r and t only.

Equation (10) is simplified by the condition that the expansion rate is isotropic and a function of cosmic time t alone. The proper distance between the points, r, θ, ϕ and $r + dr, \theta, \phi$ is

$$\ell_r = A^{1/2}\, dr\,.$$

In accordance with the usual definition, the "Hubble constant" for the local expansion rate in the radial direction is $H_r = \dot{\ell}_r/\ell_r = \dot{A}/2A$. Similarly the "Hubble constant" for expansion in the perpendicular direction is $H_\theta = \dot{B}/2B$. By isotropy these two expansion rates have to be the same, and by homogeneity the rate has to be a function of time alone,

$$\dot{A}/A = \dot{B}/B = g(t) \tag{11}$$

This equation says that B is a function of r multiplied by a function of t. By appropriate choice of the initial assignment of the radial coordinate label in some chosen initial hypersurface σ evidently we can arrange that the function of r has whatever form we like, and a convenient choice is

$$B = a(t)^2\, r^2\,. \tag{12}$$

a(t) is the expansion parameter for the model (eq. I-7). By equations (11) and (12) A must have the form

$$A = a(t)^2\, f(r)\,,$$

so the line element becomes

$$ds^2 = dt^2 - a(t)^2 f(r)\, dr^2 - a(t)^2 r^2\,(d\theta^2 + \sin^2\theta\, d\phi^2). \tag{13}$$

The form of $f(r)$ is determined by the condition that the model is homogeneous, so we should be able to move the origin of coordinates to a new spot on the surface of homogeneity and yet find that the line element is identical to (13) because the Universe looks the same.

Consider an infinitesimal coordinate transformation,

$$x^{\alpha} = \bar{x}^{\alpha} + \xi^{\alpha}(\bar{x}). \tag{14}$$

The usual coordinate transformation law is

$$\bar{g}_{\alpha\beta}(\bar{x}) = g_{\gamma\delta}(x)\frac{\partial x^{\gamma}}{\partial \bar{x}^{\alpha}}\frac{\partial x^{\delta}}{\partial \bar{x}^{\beta}}.$$

The condition of homogeneity is that, with proper choice of the ξ^{α}, the new metric tensor be identical to the old one,

$$\bar{g}_{\alpha\beta}(x) = g_{\alpha\beta}(x).$$

These last three equations collected together yield Killing's equation,

$$g_{\alpha\beta,\gamma}\,\xi^{\gamma} + g_{\alpha\gamma}\,\xi^{\gamma}_{,\beta} + g_{\beta\gamma}\,\xi^{\gamma}_{,\alpha} = 0\,.$$

With equation (13) the r and θ components of this equation are

$$f_{,r}\,\xi^{r} + 2f\,\xi^{r}_{,r} = 0,$$

$$\xi^{r} + r\,\xi^{\theta}_{,\theta} = 0,$$

$$f\,\xi^{r}_{,\theta} + r^{2}\,\xi^{\theta}_{,r} = 0.$$

The first of these equations says

$$\xi^{r} = C(\theta,\phi)\,f(r)^{-1/2}.$$

On dividing the second equation by r and then differentiating by r, and differentiating the third equation by θ, one can eliminate ξ^{θ}, to obtain

$$\frac{1}{C}\frac{d^{2}C}{d\theta^{2}}(\theta,\phi) = r^{2}\,f(r)^{-1/2}\frac{d}{dr}(f_{(r)}^{-1/2}\,r^{-1}) = D \tag{15}$$

where D evidently must be a constant. We can determine D by noting that in the neighborhood of $r = 0$ the transformation (14) should look like the ordinary law for infinitesimal translation of the origin of spherical coordinates in flat space. If the translation is along the direction of the polar axis this law is $\xi^r \propto \cos\theta$. This is consistent with (15) if $D = -1$. Then (15) gives

$$f = (1 - r^2/R^2)^{-1}.$$

Following the usual convention the constant of integration has been written as R^{-2}, although of course it can be positive or negative. With this result (13) reduces to the Robertson-Walker form,

$$ds^2 = dt^2 - \frac{a(t)^2\, dr^2}{1-r^2/R^2} - a(t)^2\, r^2(d\theta^2 + \sin^2\theta\, d\phi^2). \qquad (16)$$

iii) *Expansion Rate*

The time dependence of the expansion parameter $a(t)$ is determined by the gravity theory, which we are taking to be general relativity with $\Lambda = 0$. Then one can get the desired differential equation either by substituting the form of g_{ij} from (16) into Einstein's field equations, as given for a general diagonal metric tensor by Dingle,[7] or more simply by following the argument given in Chapter I (eq. I-10). The result in either case is the equation

$$\frac{\dot{a}^2}{a^2} = \frac{8}{3}\pi G\rho - \frac{1}{R^2 a^2}, \qquad (17)$$

where it will be recalled that ρc^2 is the total energy density of matter and radiation, including annihilation energy.

b) *Properties of the Lemaître Model*

When $R^{-2} > 0$ the space is said to be closed, the radius of curvature of space being aR. To see this consider the ratio of circumference to radius of a circle. We can choose coordinates such that the origin is at the center and the circle is the line $r = $ constant, $\theta = \pi/2$. Then by (16) the proper radius of the circle is

$$\ell = \int dr\, a/(1 - r^2/R^2)^{1/2}$$

$$= aR \sin^{-1} r/R\,,$$

and the circumference is

$$C = 2\pi\, ar\,.$$

This may be compared to a circle on a two-dimensional spherical surface. If the radius of the sphere is aR and if the circumference of the circle on the sphere is $C = 2\pi\, ar$, then the radius of the circle is ar, and this radius subtends angle

$$\psi = \sin^{-1} r/R \tag{18}$$

at the center of the sphere. Thus the radius of the circle, measured *on* the surface of the sphere, is

$$\ell = aR\,\psi = aR \sin^{-1} r/R\,,$$

as before.

A convenient coordinate transformation in equation (16) is to replace r with the angular variable ψ (eq. 18) for the closed model, and

$$\psi = \sinh^{-1} r/R \tag{19}$$

in the open model (where R^{-1} is of course the magnitude of the square root of R^{-2}). This gives

$$\begin{aligned} ds^2 &= dt^2 - a^2R^2\,[d\psi^2 + \sin^2\psi(d\theta^2 + \sin^2\theta\, d\phi^2)]\,, \\ ds^2 &= dt^2 - a^2R^2\,[d\psi^2 + \sinh^2\psi(d\theta^2 + \sin^2\theta\, d\phi^2)]\,, \end{aligned} \tag{20}$$

in the two models. The first form carries the line element past the coordinate singularity in (16) at $r = R$, $\psi = \pi/2$, to the antipodal point at $\psi = \pi$. In the second form, for the open model, the volume of space is infinite because $\sinh \psi$ increases without limit as ψ increases. Of course, this is a model result, not a statement about the real world.

Let us consider next the behavior of a free particle moving relative to the comoving observers. This might be an object shot out of a galaxy, or an intergalactic cosmic ray, or in a limiting case a photon. With the origin of coordinates chosen on the path of the particle the trajectory is $\theta = \phi =$ constant. When the particle passes a comoving observer the observed particle velocity is the coordinate velocity $v = dr/dt$ multiplied by $g_{rr}^{1/2}$. The observed momentum is then

$$p = \frac{mav}{(1 - r^2/R^2)^{1/2}} \left[1 - \frac{v^2 a^2}{1 - r^2/R^2} \right]^{-1/2} .$$

With this expression the radial component of the geodesic equations of motion (2) is

$$\frac{d}{dt} \frac{ap}{(1 - r^2/R^2)^{1/2}} = \tfrac{1}{2} avp(1 - r^2/R^2)^{1/2} \frac{\partial}{\partial r}(1 - r^2/R^2)^{-1} . \tag{21}$$

The time derivative is supposed to be taken along the path of the particle,

$$\frac{d}{dt} = \frac{\partial}{\partial t} + v\frac{\partial}{\partial r} .$$

One can use this expression to change the derivative with respect to r in (21) to a total time derivative. The result is

$$\frac{d}{dt}(pa) = 0 .$$

That is, the particle momentum varies inversely as the expansion parameter. As in the discussion of equations (I-13,15) one should not interpret this result as due to some sort of cosmic drag force on the particle — the proper momentum p observed by the comoving observer is decreasing because no matter which way the particle moves it always overtakes comoving observers that are moving away from it.

The path of a light ray ("photon") is simply determined by the condition that $ds = 0$ along the path. If the photon is propagating in the radial direction then $d\theta = d\phi = 0$ and equation (20) gives

$$dt = -Ra(t)\,d\psi\,,$$
$$\psi(t) = R^{-1}\int_t^{t_0} dt/a(t)\,. \tag{22}$$

In the cosmologically flat model, $R^{-2} = 0$, one finds from (16)

$$r(t) = \int_t^{t_0} dt/a(t)\,. \tag{23}$$

In these equations we are at the origin of coordinates, $r = \psi = 0$, at present epoch t_0, while $r(t)$, $\psi(t)$ give the coordinate position of the photon at time t.

A galaxy with zero peculiar velocity is at a fixed coordinate position r. If the galaxy emits two light pulses separated in time by δt the constancy of r and ψ in (22) and (23) demands that the pulses arrive at our position separated by the time interval

$$\delta t_0 = \delta t\, a(t_0)/a(t)\,. \tag{24}$$

This was Lemaître's original derivation of the cosmological redshift law,

$$1 + z \equiv \frac{\lambda(t_0)}{\lambda(t)} = \frac{\delta t_0}{\delta t} = \frac{a(t_0)}{a(t)}\,.$$

In the early Universe, as $t \to 0$, the expansion parameter approaches zero, but more slowly than t, so r and ψ in equations (22) and (23) converge to finite limits as $t \to 0$. That is, there is a horizon, and unless we can get by the singularity at $a = 0$ we can only survey a finite amount of space, whether the Universe is open or closed.

c) *Test of the Lemaître Model – Angular Size*

We will consider first the expected angular size of a distant galaxy. Suppose that the origin of coordinates is placed at our position, so that observed light rays move along lines of constant θ, ϕ. Two light rays leave opposite ends of the observed galaxy at epoch t_e and intersect at angle θ at our point of observation, at present epoch t_0. By equation (16) the proper diameter of the galaxy is

$$d = a(t_e)\, r\, \theta, \tag{25}$$

where the coordinate position r of the galaxy is fixed by the condition that the light ray emitted at t_e reach us at t_0; r is given by equation (23) for the cosmologically flat model, equations (18), (19) and (22) for the closed and open models. Because r and ψ approach finite limits as $t_e \to 0$, $a(t_e) \to 0$, (25) says that the angular size of a galaxy of fixed proper size d grows with increasing redshift (decreasing t_e) when the redshift is large. To see how this comes about notice that the proper distance as customarily defined between us and the distant galaxy, measured at the epoch t_e of emission of the radiation, is by (20)

$$\ell = a(t_e)\, R\psi = a(t_e) \int_{t_e}^{t_0} dt/a(t). \tag{26}$$

Again, as $t_e \to 0$ the integral approaches a constant value and $a(t_e) \to 0$, so $\ell \to 0$. That is, when the redshift is large the proper distance to the object measured at the epoch of emission *decreases* with increasing redshift. This apparently paradoxical situation arises because at first the proper distance (26) increases more rapidly than the velocity of light, so the radiation making its way in our direction at first is moving away from us. The presence of matter decreases $\dot{\ell}$, eventually making it $< c$. Because the photons move along radial lines it is evident that the light from a galaxy at very high redshift has to cover a large area in the sky, for at emission the object was looming over us.

To see the magnitude of the effect let us compute the expected angular size in the cosmologically flat Einstein-deSitter model, where

$$a(t) = a_0 (t/t_0)^{2/3} = a_0/(1+z),$$

and by (23)

$$\begin{aligned} r &= \frac{3t_0}{a_0}\left[1 - (t_e/t_0)^{1/3}\right] \\ &= \frac{2}{Ha_0}\left[1 - (1+z)^{-1/2}\right]. \end{aligned} \tag{27}$$

Equations (25) and (27) give

$$\theta = Hd(1 + z)/[2(1 - (1 + z)^{-1/2})]. \tag{28}$$

This function is minimum at redshift $z = 5/4$.

For a large galaxy the diameter d might be 20 kpc. The angular size of the galaxy at the minimum, $z = 5/4$, would be 4.6 arc seconds, at the largest observed redshift for a galaxy, $z = 0.46$, it would be 5.8 arc seconds, and at the largest redshifts of quasars, $z \sim 3$, it would be 5.5 seconds, in the Einstein-deSitter model. The best angular resolution obtained looking through the atmosphere is ~ 1 arc second, so in this model giant galaxies ought to be resolved whatever their redshift.

It is amusing to compare (28) with the expected result in a "tired light" model.[8] In a static cosmological model the optical and radio redshift observations would be consistent with the *ad hoc* law

$$d\lambda/d\ell = \lambda H,$$

for the rate of loss of energy of a photon per unit path length. Here H would not depend on time. In this model the distance of an object at redshift z is

$$\ell = H^{-1} \ln (1 + z),$$

and assuming Euclidian geometry the angular size would be

$$\theta = d/\ell = Hd/\ln (1 + z).$$

At redshift $z = 0.46$ this is smaller than (28) by a factor of 1.6.

The angular size test is difficult because astronomical objects like galaxies are fuzzy. Baum has completed an ingenious scheme for measuring angular sizes of distant galaxies, with the results shown in Figure VI-1.[9] Each data point is the mean for a number of galaxies in one rich cluster. The highest solid curve in the figure is the Einstein-deSitter model, equation (28), where the one adjustable parameter, the galaxy size d, is fitted to the low redshift points. The closed general relativity models with $\Lambda = 0$

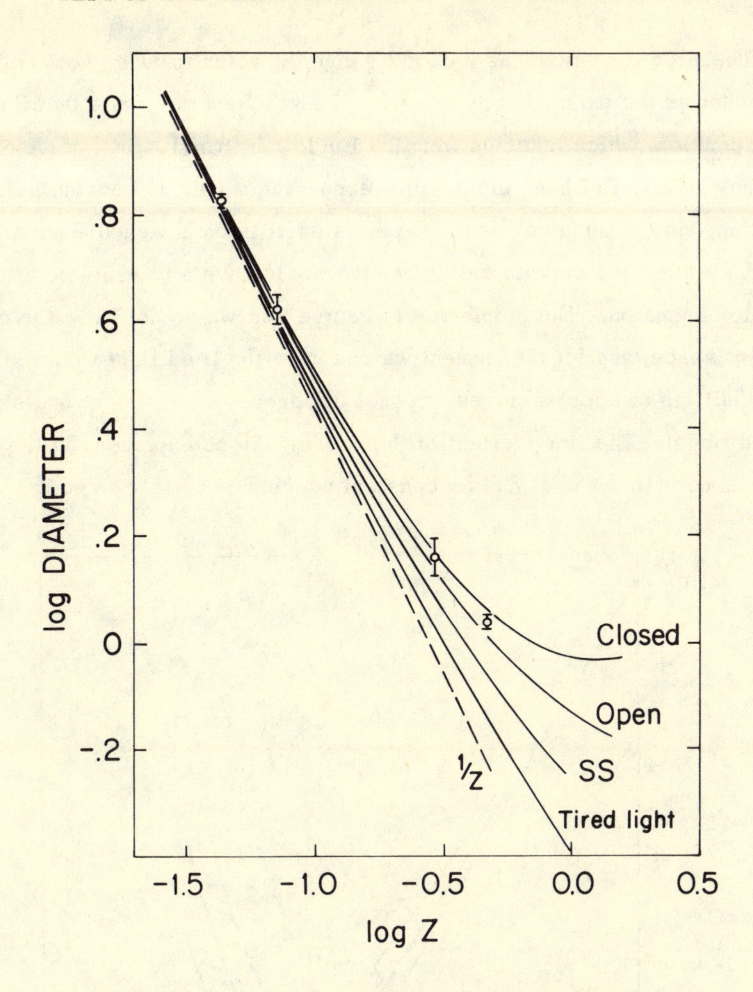

Fig. VI-1. Baum Angular Size Data. The closed Lemaître models with $\Lambda = 0$ fall above the top solid line, and the open Lemaître models with $\Lambda = 0$ fall between this line and the next one. Also shown are the predictions of the Steady State model and of the flat tired light model.

are above this curve, the open models below it and above the second highest curve. The most striking thing about the figure is that the data points do show an apparent break away from the naive formula $\theta \propto z^{-1}$. The significance of this is that one may hope to be able to use it as direct evidence of the effect of expansion of the Universe.

The tired light model is a useful gauge by which to judge how well established is the expansion hypothesis. We see from the figure that the effect of expansion, which takes us from the Euclidian "tired light" curve to the cosmologically flat Einstein-deSitter expanding model, is considerable, well beyond the apparent limits of precision of Baum's measurements. That is, if we knew the curvature of space we would have a measurable effect due to expansion. The problem is of course that we do not know the curvature of space, and for the moment we can save the tired light model simply by admitting (*ad hoc*) a curved, closed, homogeneous, isotropic and static world model. The line element of this model may be expressed in the form of equation (16), where a is a constant which we can take to be unity, and

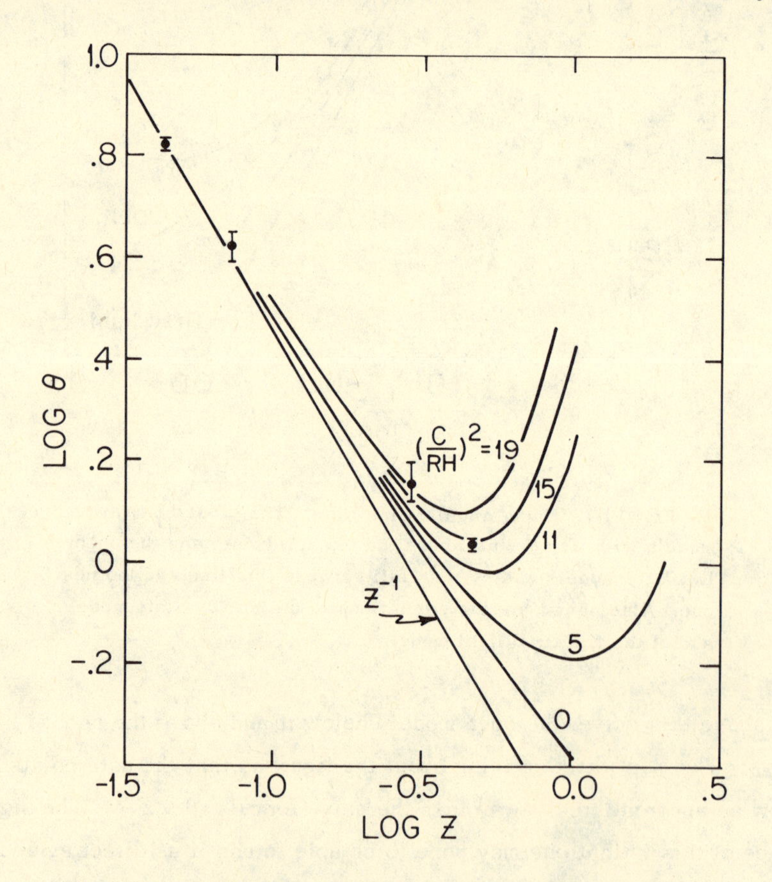

Fig. VI-2. Angular Size in Curved Tired Light Cosmologies.

we find that the relation between measured angular size and proper linear size d of a galaxy at redshift z is, in this curved tired light model,[8]

$$\theta = \frac{d}{R \sin [(\ln(1+z))/HR]} .$$

The plot of angular size as a function of z assuming fixed linear size d independent of redshift is shown in Figure VI-2. It is apparent that the data are fitted by the parameter choice

$$\left(\frac{1}{HR}\right)^2 \sim 15$$

or

$$R \sim 800\, h^{-1}\, \mathrm{Mpc} .$$

The effect on the angular size of a distant galaxy due to the irregular distribution of the intervening matter has been discussed by a number of authors, with varying conclusions.[10] Since this problem has not yet been worked out in all detail we present here only a rough estimate of the orders of magnitude, as follows. Suppose a galaxy with proper diameter d is observed at redshift z, and suppose that at redshift $z_1 < z$ the light from the galaxy is deflected by some density irregularity, so that the ray from one edge of the galaxy is bent through an angle $\delta\theta_1$ relative to the ray from the other edge. Let r be the coordinate position of the observed galaxy when our position is at the origin, and let r_1 be the coordinate position of the observed galaxy when the perturbation is at the origin, the coordinates being defined by equation (16). Then in the absence of the perturbation the observed angular size of the galaxy would have satisfied

$$d = ar\,\theta .$$

The observed angular size $\theta + \delta\theta_0$ makes the apparent size of the galaxy

$$d' = ar\,(\theta + \delta\theta_0) .$$

But since

$$d' - d = ar_1\, \delta\theta_1 ,$$

we have[10]

$$\delta\theta_0 = \frac{r_1}{r}\,\delta\theta_1 .$$

In the Einstein-deSitter model this becomes (eq. 27)

$$\delta\theta_0 = \delta\theta_1 \frac{[(1+z_1)^{-1/2} - (1+z)^{-1/2}]}{[1-(1+z)^{-1/2}]} .$$

If the perturbation were a large galaxy, mass $10^{11}\,\mathfrak{M}_\odot$, the deflection at $R = 10$ kpc impact parameter would be

$$\delta\theta_1 = \frac{4\,G\mathfrak{M}}{Rc^2} \sim 0.4'' .$$

If the perturbation were a great cluster of galaxies the mass might be larger by four orders of magnitude, the radius larger by two orders of magnitude, making the deflection angle two orders of magnitude larger. However, what is relevant here is the change in convergence of rays from either side of the observed galaxy. If the observed galaxy has a diameter of 10 kpc the resulting $\delta\theta_1$ would be reduced by about two orders of magnitude, making the result about comparable to what we got for the single large galaxy.

The deflection $\delta\theta_1 \sim 0.4''$ is appreciable, but it is unlikely that the line of sight would pass close enough to a galaxy for this to happen in any randomly chosen case. If the observed galaxy is large, $\mathfrak{M} \sim 10^{11}\,\mathfrak{M}_\odot$, the relevant number is the probability that another large galaxy will be centered roughly within the solid angle Ω subtended by the observed galaxy. For $z < 1$ this is

$$P \cong \frac{\Omega}{3}\left(\frac{z}{H}\right)^3 n$$

$$\cong 0.01\,z ,$$

where for the number density n we have used equation IV-3. Since P is small the typical case will be that the ray suffers a number of small perturbations from more distant galaxies.

The mean distance between big galaxies is roughly 2 Mpc. With the impact parameter at 1 Mpc the scattering angle $\delta\theta_1$ is reduced by a factor ~ 100 because the radius in the denominator is that much larger, and one loses an extra factor $\sim$ (1000 kpc)/(10 kpc) in figuring the perturbation to the convergence of the light rays, assuming they are 10 kpc apart, giving $\delta\theta_1 \sim 10^{-4''}$. If $z = 1$ the light ray passes $\sim$ 1000 large galaxies at ~ 1 Mpc, the random sum yielding $\delta\theta_0 \sim 10^{-2''}$, which is uninterestingly small. The more detailed calculation of Gunn gave about 1 percent perturbation in angular size due to large galaxies.[9] There is of course the possibility that the calculation is in error because it ignores some class of objects, but for the moment the tentative indication is that this effect is not large enough to affect the statistical test.

d) *Redshift-Magnitude Test*

The next important test is the relation between apparent magnitude and redshift of a distant galaxy. In the expanding model the calculation goes as follows. It is convenient this time to center the coordinates on the observed galaxy. Let $\mathfrak{L}$ be the total luminosity of the galaxy (ergs sec^{-1}) integrated over all frequencies. This radiation reaches us at coordinate position r spread over the surface area $4\pi a_0^2 r^2$. The energy of each photon is diminished by the redshift factor $(1 + z)$, and the rate at which the photons arrive is diminished by this same factor (eq. 24). Thus the observed total energy flux (ergs sec^{-1} cm^{-2}) from the galaxy is

$$f = \mathfrak{L}/[4\pi a_0^2 r^2 (1 + z)^2]. \tag{29}$$

If the surface brightness of the galaxy were constant over the radiating disc the observed surface brightness would be the energy flux (29) divided by the solid angle Ω subtended by the galaxy. For a spherical galaxy

$$\Omega = \pi\theta^2/4. \tag{30}$$

By equations (25), (29) and (30), the observed surface brightness (ergs cm^{-2} sec^{-1} ster^{-1}) is

$$i = \frac{f}{\Omega} = \left(\frac{\mathcal{L}}{\pi^2 d^2}\right)(1+z)^{-4} . \tag{31}$$

The first factor on the right hand side is the ordinary expression for the surface brightness of the uniformly bright galaxy as determined by a nearby observer. This brightness is reduced by the fourth power of the redshift factor, independent of cosmological model. An earlier example of this same effect was that the Primeval Fireball temperature T is proportional to $(1+z)^{-1}$, so the brightness $\propto T^4 \propto (1+z)^{-4}$.

Sandage has directed considerable effort to the relation between apparent magnitude and redshift.[1] Using (29) corrected to observation in a fixed bandwidth he determines the acceleration (or deceleration) parameter

$$q = -(\ddot{a}a/\dot{a}^2)_0 \cdot \tag{32}$$

This is a fixed parameter, the acceleration in the *present* Universe. q is proportional to the second-order term in a Taylor series expansion of $a(t)$ about the present epoch, the first-order term being Hubble's constant. Details of how this parameter may be used to compute the expected behavior of the curves of apparent magnitude or of angular size as functions of redshift are given in references (2) and (3).

In the general relativity cosmologies with $\Lambda = P = 0$ we have from equations (I-10,11)

$$2q = 1 + \frac{R^{-2}}{a_0^{\ 2} H^2} , \tag{33}$$

which means $q = \frac{1}{2}$ in the Einstein-deSitter model, $q > \frac{1}{2}$ in the closed models $R^{-2} > 0$. Sandage's results to date show no appreciable deviation from the naive law $f \propto z^{-2}$ to redshift $z \sim 0.2$, and are consistent also with acceleration parameter in the neighborhood of ½. It is too soon to say whether this test favors an open or closed model.

In the tired light model the net energy flux from a distant galaxy with luminosity $\mathcal{L}$ would be[8]

$$f = \frac{\mathcal{L}}{4\pi R^2 (1+z) \sin^2 [(\log (1+z))/HR]^2}, \tag{34}$$

where again R is the radius of curvature of the homogeneous static space. It will be noted that there is only one power of $(1+z)$ in the denominator, as opposed to two powers of $(1+z)$ in the denominator of (29), because here the photons are supposed to be losing energy, but the rate of receiving photons is not affected by the tired light process if photons are conserved. Hubble called these two effects the "energy effect" and the "number effect;" the point is that if the number effect could be isolated and measured it could be considered direct evidence of expansion.[8,11]

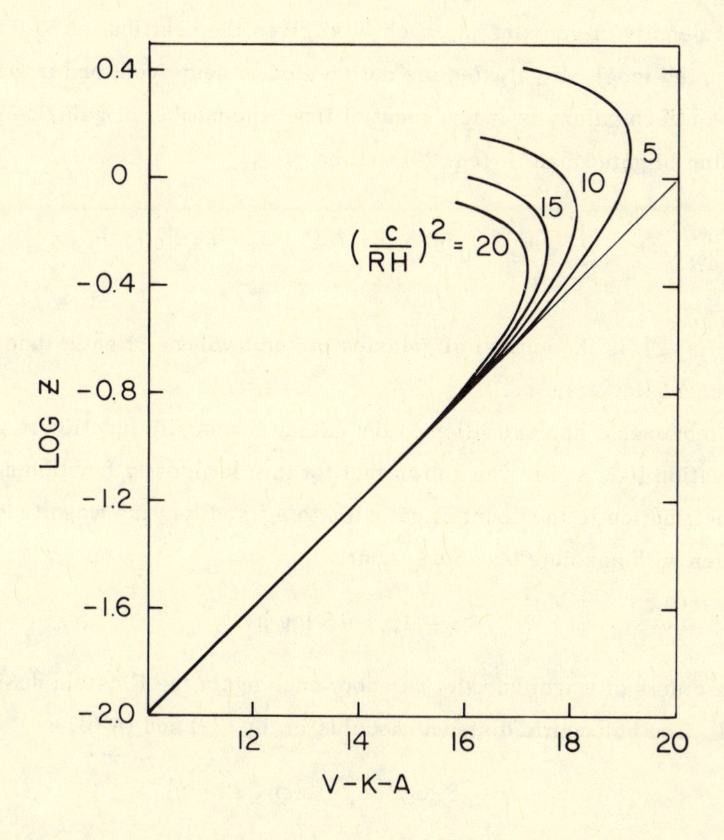

Fig. VI-3. Redshift-Magnitude Relation in Tired Light Cosmologies.

The bolometric magnitude from (34) is plotted as a function of z in Figure VI-3. As Sandage's data extending to $z \sim 0.2$, and corrected to apparent bolometric magnitude, show no deviation from linearity, it appears that the parameter $1/RH$ could not exceed ~ 4, apparently just excluding what would be needed to fit Baum's angular size measurements.

e) *Hubble's Test*

Hubble's cosmological test was to count galaxies as a function of limiting apparent magnitude. Suppose we observe now a light ray that at epoch t was at coordinate distance $r(t)$. If we look in solid angle Ω we are looking over an area $\Omega\, a(t)^2\, r(t)^2$ (eq. 16) at t. This area multiplied by dt is the proper volume swept out in time dt, and when multiplied by the number density of galaxies at epoch t it gives the contribution to the number counted in Ω. If galaxies are not created or destroyed, and if the luminosity of each galaxy is independent of time, the number of galaxies per steradian brighter than f (eq. 29) will be

$$\frac{dN}{d\Omega}(>f) = \int_0^{t_0} dt\, r^2 a_0{}^3 a(t)^{-1}\, n\,(>\mathcal{L} = 4\pi\, r^2 f\, a_0{}^4 a(t)^{-2}), \tag{35}$$

where $n(>\mathcal{L})$ is the number of galaxies per unit volume brighter than $\mathcal{L}$ in the present Universe.

A reasonable approximation to the galaxy luminosity function is given by equation II-1, and it was shown that for this luminosity function most of the contribution to the count of galaxies to a fixed limiting magnitude is by galaxies with absolute magnitude near

$$M_V^* \cong -19.5 + 5 \log h\,. \tag{36}$$

To fix orders of magnitude, let us adopt once again the Einstein-deSitter model. The bolometric distance modulus is, by (29) and (A-6),

$$m - M = 5 \log\,[r a_0{}^2/(a(t) \times 10\ \text{pc})]\,,$$

and we have from (27) and (36)

$$M(z,m) - M^* = m - 22.9 - 5 \log 2(1+z)\,[1 - (1+z)^{-1/2}]. \qquad (37)$$

When we set $M = M^*$ this equation gives us an estimate of the typical redshift of galaxies observed at apparent magnitude m. The result is listed in Table 1.

TABLE VI-1

REDSHIFT AND APPARENT MAGNITUDE

z	m	K(1) = 2.5 log (1 + z)	$K_R(2)$	m_R
0.02	14.4	0.02	0	14.1
0.1	18.0	0.10	0	17.8
0.2	19.5	0.20	0.07	19.5
0.28	20.3	0.27	0.12	20.4
0.5	21.6	0.44	---	---
1.0	23.3	0.76	---	---
2.0	24.9	1.19	---	---

It will be recalled that m is a bolometric magnitude. Since observations are made in a fixed bandwidth the magnitude must be corrected for the effect of the redshift of the spectrum. The energy flux observed in a narrow bandwidth $\Delta\lambda$ at wavelength λ is

$$f_\lambda\,\Delta\lambda = \left[\frac{\mathcal{L}(\lambda/(1+z))}{4\pi a_0^2 r^2 (1+z)^2}\right]\frac{\Delta\lambda}{1+z},$$

where $\mathcal{L}(\lambda)$ is the luminosity per wavelength interval. The first factor is obtained in the same way as equation (29), and the second factor is just the bandwidth shifted back to the source. In the red $\mathcal{L}(\lambda)$ for galaxies is nearly flat, independent of wavelength, so the correction factor to a fixed bandwidth is just the bandwidth factor $(1+z)^{-1}$, or (ref. IV-7)

$$\Delta m \equiv K(1) = 2.5 \log (1+z). \qquad (38)$$

This correction is listed in Table 1. There is a correction K(2) which depends on the departure from a flat spectrum. The fourth column in the table is the correction factor derived by Oke and Sandage for observations in the red, centered on 6300 A (ref. IV-8). The fifth column shows the apparent red magnitude,

$$m_R = m - 0.3 + K(1) + K_R(2) , \qquad (39)$$

where the term 0.3 is the difference between visual and red magnitudes of nearby galaxies.[3]

To see how the galaxy counts should deviate from the naive law $n \propto f^{-3/2}$, let us work as an example the Einstein-deSitter model with the luminosity function (II-1). Equation (35) with (27) for this model gives

$$N(<m) \propto \int_1^\infty \frac{d\omega}{\omega^{3/2}} (1 - \omega^{-1/2})^2 \, n(<M(m,z)) , \qquad (40)$$

where $\omega = 1 + z$. We will assume that the spectrum $\mathcal{L}_\lambda$ is flat, so $M(m, z)$ is given by equations (37) and (39) with $K(2) = 0$. In computing (40) the integral is cut off 3 magnitudes brighter than M^*. The interesting quantity is the ratio of the galaxy counts $N(<m)$ in the model to what would be expected according to the simple law

$$N_0 \propto 10^{0.6m} ,$$

with the ratio normalized to unity at low redshift. This ratio is listed in Table 2. For the numbers in parentheses the color correction K(2), which

TABLE VI-2

GALAXY COUNTS

m_R	N/N_0
16	0.71
18	0.49
20	(0.26)
22	(0.10)

we have ignored, becomes important, so the numbers should not be taken too seriously. This correction could of course be taken into account in some approximation in a more detailed calculation.

The galaxy count test is particularly attractive because the measurement of apparent magnitude is much easier, and can be carried much deeper, than the measurement of redshift or angular size, so the departure from the naive estimate can be much larger. Apparent magnitude $m_R = 20$ is well within reach, the K correction is not yet serious, yet the effect is a factor of about 4 in the Einstein-deSitter model. The test could be frustrated if there are many more faint galaxies than implied by (II-1), but despite this question it seems to be particularly promising for future research.

The Hubble test has been very extensively applied to the counts of radio sources.[12] The great problem in using these data as a cosmological test has been that one does not know the luminosity function of the sources, and one does not know ahead of time how the luminosity function has varied with time.

f) *Numerical Results for the Lemaître Model*

With the discovery of quasars with large and apparently cosmological redshifts, and with the prospect for observations of galaxies at redshifts approaching unity, astronomers are finding some need for an easy conversion table from observed to intrinsic quantities. This section contains a list of formulas and graphs for this purpose.

It will be assumed here that pressure and the Cosmological Constant both may be ignored. Then the parameters in the model are the acceleration parameter q (eq. 32) and Hubble's constant, which we write as usual as

$$H = 100\ h\ \mathrm{km\ sec^{-1}\ Mpc^{-1}}\ .$$

The expansion rate (eq. 17) may be rewritten with the help of (33) as

$$\frac{\dot{a}(t)}{a(t)} = H(1+z)\,(2q\,z+1)^{1/2}\,, \tag{41}$$

and the rate of change of coordinate position for a photon propagating toward us along the radial direction is (eq. 16)

$$\frac{dr}{dt} = -(1 + Kr^2)^{1/2}/a(t)\,, \tag{42}$$

where by (33)

$$K = H^2(1 - 2q)\,,$$

and where we have set $a_0 \equiv 1$. Mattig has shown that the solution to equations (41) and (42) is[13]

$$r = [qz + (q-1)((1+2qz)^{1/2} - 1)]/(q^2 H(1+z))\,. \tag{43}$$

With this result and equation (29) we can write the distance modulus corresponding to redshift z as

$$m - M = 5 \log z + K(2) + K(3)\,. \tag{44}$$

Here m and M are apparent and absolute magnitude at some fixed and given wavelength or color, like $m_V - M_V$, and m is of course supposed to be corrected for absorption. The first term on the right hand side is the usual z dependence observed at low redshift, and the other two terms contain correction factors important at higher redshift. The first correction factor $K(2)$ takes account of the fact that m is measured at a fixed wavelength band, and the cosmological redshift brings different parts of the spectrum into this band at different redshifts. If the bandwidth of observation were sharp $K(2)$ would be simply

$$K(2) = 2.5 \log\left[\frac{f(\lambda)}{f(\lambda/(1+z))}\right], \tag{45}$$

where f is the observed incident energy flux per wavelength increment (corrected for absorption) and λ is the wavelength of observation. Sandage's K term is (ref. IV-8)

$$K = K(2) + K(1)\,,$$

where $K(1)$ is defined by equation (38). We have preferred to lump $K(1)$ in with the relativistic corrections in the definition of $K(3)$,

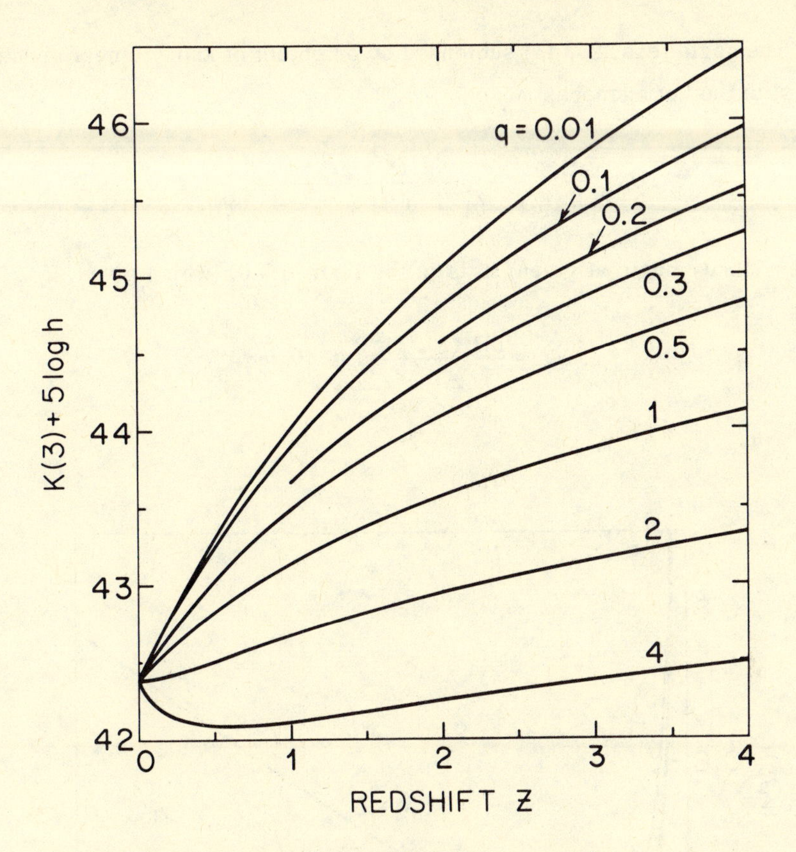

Fig. VI-4. Redshift-Magnitude Relation in Lemaître Models. K(3) is the relativistic correction factor in equation (44).

$$K(3) = A + 5 \log \left\{ \frac{(1+z)^{1/2}}{q^2 z} \left[qz + (q-1) \left((1+2qz)^{1/2} - 1 \right) \right] \right\}.$$

K(3) is independent of color of observation. The constant A fixes the scale,

$$A = 5 \log \frac{c}{H(10\ \mathrm{pc})} = 42.38 - 5 \log h\,.$$

K(3) is plotted as a function of z and q in Figure VI-4. It will be noted that the departure from the simple law $m - M = 5 \log z + \text{constant}$ is not all that large, not more than a factor of about 10 in luminosity for $z = 3$.

The angle (eqs. 25, 43) subtended by an object of known linear diameter d is, in the homogeneous model,

$$\theta = \frac{dH}{c} \frac{q^2 (1+z)^2}{[qz + (q-1)((1+2qz)^{1/2} - 1)]} .$$

The minimum angle at given z is in the limit $q \to 0$, which gives

$$\theta = \frac{dH}{c} \frac{(1+z)^2}{z + z^2/2}, \; q = 0 .$$

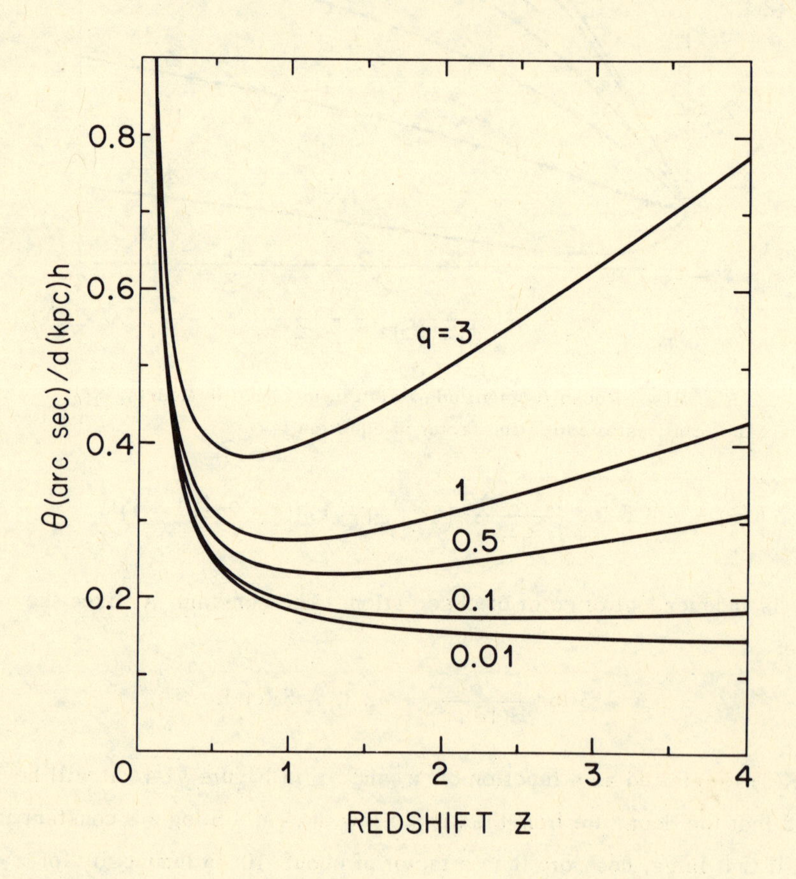

Fig. VI-5. Angular Size in Lemaître Models.

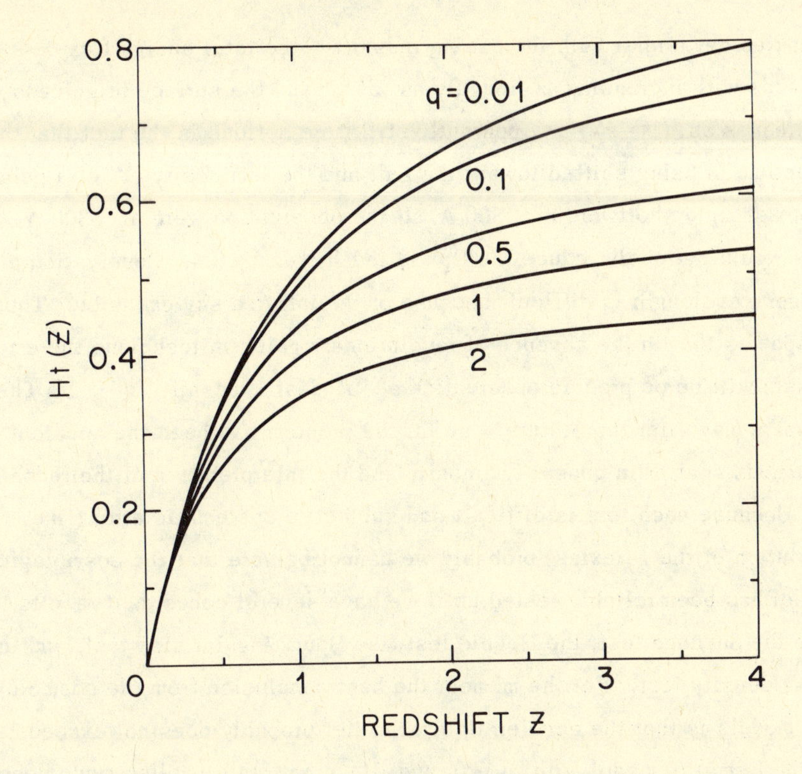

Fig. VI-6. Ages of Observed Objects in Lemaître Models.

The angular diameter is given as a function of z and q in Figure VI-5. If the diameter d is measured in kpc the observed angular diameter in seconds of arc is the quantity plotted in the figure multiplied by d and multiplied by $h = H/(100 \text{ km sec}^{-1} \text{ Mpc}^{-1})$.

Finally, it is sometimes amusing to know what is the age of a distant object as we now observe it. This is shown in Figure VI-6. The convenient dimensionless number is $Ht(z)$, where $t(z)$ is the time measured from the present since the light now observed left the object (at redshift z).

g) *Conclusions*

We would very much like to have observations to cosmological redshift $z \sim 1$ and beyond, first simply to check that the Universe looks about as we expect, second because the expansion effects should become substantial and even manifest at high redshift. The first problem with this goal is that galaxies get fainter with increasing redshift – the total energy flux decreases with increasing z roughly as z^{-2}, and the surface brightness decreases as $(1+z)^{-4}$. Apparently still more serious is the fact that the spectrum is being shifted toward the red, and the luminosity $\mathfrak{L}$ of a galaxy drops sharply shortward of 4500 A. If the observation were at 6300 A this would seriously reduce the flux at redshift $z \gtrsim 0.4$. Observation at longer wavelength is difficult because of the infrared sky emission. Thus it appears that in the absence of revolutionary gains in technique the emphasis will be on precise observation at "modest" redshift, $z \lesssim 1$. The quasars may alter the picture – so far the problem has been the apparent enormous scatter in quasar luminosity and the interpretation of the redshift.

Because each test is difficult and subject to systematic error (like evolution of the galaxies) probably we cannot be sure that the cosmological model has been reliably tested until we have several concordant results, like the Sandage test, the Hubble test, the Baum angular size test, and the mass density test. For the moment the best conclusion from the cosmological models is that the acceleration parameter probably does not exceed 1.5. In the Lemaître model with $\Lambda = 0$ and $P << \rho c^2$ this implies mass density

$$\rho_0 \lesssim 6 \times 10^{-29}\, h^2 \text{ g cm}^{-3},$$

which certainly is not ruled out by the available observations described in Chapter IV. If the Universe were closed the present radius of curvature would satisfy

$$a_0 R = (2q-1)^{-1/2}\, cH^{-1} \gtrsim 2000\, h^{-1} \text{ Mpc}.$$

REFERENCES

1. A. R. Sandage, *Physics Today,* February 1970, p. 34 and earlier references therein.
2. R. C. Tolman, *Relativity, Thermodynamics and Cosmology,* 1934; L. Landau and E. Lifshitz, *Classical Theory of Fields,* 1951; G. McVittie, *General Relativity and Cosmology,* 1956; H. P. Robertson and T. W. Noonan, *Relativity and Cosmology,* 1968.

3. A. R. Sandage, *Ap. J.* 133, 355, 1961.
4. The coordinate transformations are spelled out in P. J. E. Peebles, *Am. J. Phys.* 37, 410, 1969.
5. C. W. Misner, *Ap. J.* 151, 431, 1968; *Phys. Rev. Letters* 22, 1071, 1969; S. Hawking, *M.N.* 142, 129, 1969.
6. H. P. Robertson, *Proc. Nat. Acad. Sci.* 15, 822, 1929; A. G. Walker, *Q. J. Math* (Oxford) 6, 81, 1935.
7. H. Dingle, Proc. *Nat. Acad. Sci.* 19, 559, 1933.
8. This model was originally discussed by E. Hubble and R. C. Tolman, *Ap. J.* 82, 307, 1935; the modern prospects for an observational test are discussed by M. Geller and P. J. E. Peebles, 1971.
9. W. A. Baum, to be published in the proceedings of IAU Conference #44, Uppsala, 1970.
10. Ya. B. Zeldovich, *Astron. Zh.* 41, 19, 1964; Engl. tr. in *Soviet Astronomy -A. J.* 8, 13, 1964; B. Bertotti, *Proc. Roy. Soc.* A294, 195, 1966; J. E. Gunn, *Ap. J.* 147, 61, 1967.
11. Ref. I-1, Chapter 8.
12. A review is given by M. Ryle, *Ann. Rev. Astron. Astrophys.* 6, 249, 1968.
13. W. Mattig, *A. N.* 284, 109, 1958.

VII. HISTORY OF THE UNIVERSE – SCENARIOS

The previous chapters have been devoted mainly to what the Universe is like now or at modest redshift, $z \lesssim 0.3$. A second great line of attack in cosmology has been the attempt to deduce from theoretical and philosophical considerations and whatever fossil evidence we may manage to find what the Universe was like in the distant past, and perhaps thereby to come to some understanding of why the Universe is now the way it is. Understandably progress here is even more uncertain than in the relatively more modest goal of exploring the present Universe, but we do have a few memorable questions and lines of attack.

The microwave background radiation seems to be a major fossil find, the closely thermal nature of the spectrum at $\lambda \gtrsim 0.3$ cm suggesting that the Universe did expand from a dense state because there appears to be no natural way by which this spectrum could have been produced in the present Universe. On the other hand to the extent that the radiation has a thermal spectrum it is not specific about the particular interaction by which it was thermalized, or even the temperature at which it happened. The best that can be done is to ask under what conditions, what possible histories of the Universe, we would expect to find a closely thermal Fireball spectrum, and under what conditions we would not.[1]

A second major clue is the present state and organization of matter. We know some matter is lumped in galaxies, there is speculation that an undetermined part may be in more or less uniformly distributed intergalactic gas, and of course we must be prepared to encounter yet other forms. Two obvious questions are, what were the processes attending the division of some or all the matter into galaxies, and what were the processes that determined the present state of the remaining intergalactic matter? As the intergalactic medium has thus far proved a will-o'-the-wisp discussions of how it came about tend to be highly conditional. There have been a number of attempts to spell out self-consistent histories by which there could be a dense intergalactic gas ($n_0 \sim 10^{-5}$ cm^{-3}) that would avoid manifesting itself in the tests described in Chapter IV.[2] Another point of concern with a dense intergalactic medium is the gravitational instability effect, the tendency of density irregularities to grow ever more pronounced as the Universe expands.

The magnetic field of the Galaxy may be another fossil. It has been argued[3] that it is hard to see how the present galactic field, $\sim 10^{-6}$ Gauss, could have originated within the Galaxy, although this is a subject of debate.[4] If the galactic field is primeval, one might guess that the present intergalactic field would be on the order of 10^{-10} Gauss, because if matter threaded with this field, with present mean density $\sim 10^{-30}$ g cm^{-3}, were squeezed to the density within the Galaxy, $\sim 10^{-24}$ g cm^{-3}, conserving flux, it would produce a magnetic field of about 10^{-6} Gauss. It is readily verified that the lifetime against ohmic dissipation of a magnetic field with dimension comparable to the Galaxy is very much longer than H^{-1}, and that on this ground the magnetic field could have existed as a primeval feature of the Big Bang. It would be of considerable interest to understand how a primeval magnetic field might affect the course of evolution of the Universe, both near the singularity in the Lemaître model[5] and in the formation of structures like galaxies. A second even harder problem is to understand what the fact of the existence of a primeval magnetic field would be telling us about the origin of the Universe.

It has been argued that, in view of the apparent microscopic symmetry of matter and antimatter, it would be surprising if the strong asymmetry favoring matter over antimatter in the Solar System prevailed over the whole Universe.[6] It could be, for example, that the Andromeda Nebula is made of antimatter instead of matter and we would be hard put to know it until neutrino astronomy makes very substantial progress. To explain the manifest separation of matter and antimatter one might simply assume that different patches of space started out with excess matter or antimatter.[7] There have been some attempts also to find mechanisms of separation of matter and antimatter, given that the two are well mixed to begin with.[6]

Returning to the Primeval Fireball hypothesis we have the question, why should the present temperature be $T_0 = 2.7$ K? A nearly epoch-independent way to state this question is to look at the entropy. The radiation entropy per nucleon divided by Boltzmann's constant is the dimensionless number (eq. V-6)

$$s/k \sim 10^8 ,$$

which is nowhere near the other magic number of cosmology,

$$e^2/Gm_p^2 \sim 10^{40} .$$

Another way to state this is that s/k is the number of radiation quanta per nucleon. The large value of s/k may mean that in the very early, very dense Universe the expansion was highly irreversible, and that s/k is the total amount of entropy produced in the expansion. Another thought has been that this is entropy left over from an earlier cycle in an oscillating Universe. It will be noted, however, that if we accept the ordinary law that entropy cannot decrease, and if we assume the Universe is homogeneous, then we must conclude that the Universe could not have been oscillating for an infinite length of time because this would produce infinite entropy. This is the analogue in an oscillating cosmology of the much older argument that if the Universe were homogeneous and conserved energy the brightness of the sky would be as bright as the surface of stars. This very important point, which according to Jaki[8] was discovered by Halley and by Chéseaux, is called Olbers' paradox. Finally, it has been proposed by Omnes[6] that the dimensionless number s/k might be derived from first principles on the assumption that in the very early dense Universe matter and antimatter were present in equal amount, well mixed and cold, and that the Primeval Fireball radiation is the thermalized annihilation radiation produced as some of the matter and antimatter collected in separate patches, the rest annihilated.

None of these subjects has been worked out in all detail, and in some cases we have at most a schematic outline of an idea, which is not surprising in view of all the uncertainties. The present chapter is meant to be a sampler of these topics. It should be clear that we are not at all sure even what questions we would like to ask, which concepts will prove to be of lasting interest, and the selection in the sampler accordingly is largely a matter of taste. The first topic is the rate of relaxation of radiation to thermal equilibrium. We observe now a radiation background with an approximately thermal spectrum. Assuming this identification is valid, we have a constraint on possible histories of the Universe from the conditions (1) that at some point the Universe ought to have been in a state where the radiation could relax to thermal equilibrium, and (2) that since then non-thermal processes should not have seriously messed up the spectrum. The first of these points is dealt with in section (a) below, where we attempt to list the processes thought to be important for thermal relaxation of the Primeval Fireball radiation under the assumption that the Universe has been reasonably homogeneous and isotropic. The result is a characteristic relaxation time as a function of the expansion parameter a (Fig. 1 below), and the constraint is that at some point the characteristic expansion time of the Universe should have exceeded the relaxation time to guarantee a blackbody Fireball spectrum. In section (b) we consider some aspects of the fact that the Universe manifestly is not homogeneous and isotropic. The goal is to understand the evolution of irregularities in an evolving Universe. In section (c) we consider possible perturbations to the Fireball spectrum due to non-thermal processes.

a) *Interaction of Matter and the Primeval Fireball*

If the Primeval Fireball exists it must have played a central dynamic and thermal role in the early Universe because the equivalent mass density of the radiation increases with increasing redshift faster than the density of ordinary matter, nucleons and electrons. With $T_0 = 2.7$ K, at redshift $z \sim 10000$ the radiation mass density would be the largest contribution to the total, and the Universe would be radiation-dominated. Apparently by coincidence, at roughly the same redshift the radiation would be hot enough to thermally decompose hydrogen, and Thomson scattering would very strongly couple matter and radiation. The following discussion concentrates on this interesting epoch, redshift $\sim$ 1000 - 10000, temperature 3000 to 30000 K, matter density $\sim 10^3$ to 10^6 protons cm^{-3}, where matter and radiation may decouple.

The following assumptions will be adopted.

(1) There is now a Primeval Fireball, spectrum close to thermal, present temperature $T_0 = 2.7$ K.

(2) The Universe has been homogeneous and isotropic to good approximation, and the matter was uniformly distributed when it was last in thermal contact with the radiation. By the symmetry arguments leading up to equations (I-14) and (V-4) we know then that the radiation temperature and matter density at epoch z satisfy

$$T = T_0 (1 + z), \qquad n = n_0 (1 + z)^3 . \tag{1}$$

(3) The parameter n_0 in (1) may be written as (eq. IV-1)

$$n_0 = \Omega n_c = 1.12 \times 10^{-5} \Omega \, h^2 \text{ cm}^{-3}, \tag{2}$$

where the dimensionless parameter Ω is assumed to be in the range (cf. eq. IV-7)

$$0.01 \lesssim \Omega \lesssim 1 . \tag{3}$$

The thought is that at high redshift, $z \gtrsim 30$, say, matter now in galaxies along with any intergalactic matter would have been uniformly distributed gas. The choice of redshift is simply that at $z \gtrsim 30$ galaxies as we now observe them would overlap. Of course (3) is not the only possibility — for example, it has been suggested that galaxies issued from nuclei, perhaps as some sort of delayed remnants of the Big Bang.

(4) At $z \gtrsim 30$ there are no elements heavier than hydrogen. Helium may well have been present but it would not change the qualitative picture.

Because atomic hydrogen interacts only weakly with radiation the first interesting question is the redshift at which one might expect to find the matter ionized.

i) *Thermal Ionization of Hydrogen*

The equilibrium ionization of hydrogen in blackbody radiation at temperature T may be found as follows. Suppose a definite amount of matter is confined to a box of volume V at temperature T. Then the *a priori* probability assigned to a definite state of the matter, with energy E, is proportional to

$$e^{-E/kT} . \tag{4}$$

Now the problem reduces to a counting game – one must sum (4) over that subset of states all of which have some chosen ionization of the hydrogen, and then seek that ionization that makes the sum S a maximum. To good accuracy the energy E in (4) can be written as a sum of the free single particle energies of the electrons, protons, and hydrogen atoms, so the sum over (4) looks like

$$S = \sum e^{-E_{n_1}} e^{-E_{n_2}} \ldots , \tag{5}$$

where each of the E_n represents the energy of a single electron, or proton or atom. The sum is taken over all choices of the n_i for N_e electrons, N_e protons, and $N\text{-}N_e$ atoms. Because densities are low, the probability is small that any chosen single particle state will be occupied, so we make little error by adding to the sum (5) the forbidden combinations where two or more identical particles occupy the same single particle state. When these terms are added (5) reduces to the form

$$S = \frac{Z_e^{N_e} Z_p^{N_e} Z_H^{N-N_e}}{N_e!^2 \, (N-N_e)!} . \tag{6}$$

Here

$$Z_e = \sum e^{-E_n/kT} , \tag{7}$$

where the sum is over all single electron states n in V. The other two Z's are similarily defined. The factorials are introduced in the denominator of (6) because $Z_e^{N_e}$ generates a given set of electron quantum numbers, $(n_1, n_2...)$, $N_e!$ times when the quantum numbers are all different, which is nearly all the time, and we only want to count a given set once. The next step is to change the sum (7) to an integral. The number of single particle states in V with momentum between p and $p + dp$ is

$$2(4\pi\, p^2 V\, dp) / h^3 ,$$

where the factor of 2 takes account of the two spin states of the electron (and Planck's constant will not be confused with the dimensionless form of Hubble's constant). With this formula for the density of states, and the energy equations $E_e = p^2{}_e/2m_e$, $E_H = p^2{}_H/2m_H - B$, where B is the hydrogen atom binding energy, we have

$$\begin{aligned} Z_e &= 2(2\pi\, m_e kT)^{3/2}\, V/h^3 , \\ Z_H &= 4(2\pi\, m_H\, kT)^{3/2}\, Ve^{B/kT}/h^3 , \end{aligned} \tag{8}$$

and similarly for the protons. To find the value of N_e that makes S maximum the standard trick is to take the logarithm of (6), use Stirling's approximation

$$\ln N! = N \ln N - N ,$$

and then require

$$\frac{\partial}{\partial N_e} \ln S = 0 .$$

The result is the Saha formula,

$$\frac{N_e^2}{N-N_e} = \frac{Z_e\, Z_p}{Z_H} = \frac{(2\pi\, m_e\, kT)^{3/2}}{h^3}\, Ve^{-B/kT} .$$

The fractional ionization x is defined as

$$x = N_e/N,$$

where N is the number of atoms plus free protons, and x evidently satisfies

$$\frac{x^2}{1-x} = \frac{(2\pi m_e kT)^{3/2} e^{-B/kT}}{n h^3}, \tag{9}$$

with $n = N/V$ the nucleon number density.

The nucleon density and the temperature are given as functions of redshift by equations (1)-(3). When these quantities are substituted in (9) one finds the equilibrium fractional ionization listed in Table VII-1. It is apparent that the equilibrium favors ionization at redshift greater than about 1400, with only slight variation over the assumed reasonable range of Ωh^2. The rate of relaxation to this equilibrium is considered in the next section.

TABLE VII-1.

EQUILIBRIUM FRACTIONAL IONIZATION

Redshift $1+z$	Temperature T	Density Parameter Ωh^2			
		10	1	0.1	0.01
1000	2700 K	0.00001	0.00004	0.00011	0.0003
1200	2970	0.00125	0.0040	0.0124	0.0388
1400	3240	0.0356	0.108	0.303	0.664
1600	3780	0.358	0.732	0.954	0.995
1800	4860	0.914	0.990	1.	1.
2000	5400	0.995	0.9995	1.	1.

ii) *Thermal Ionization and Recombination Rates*

The rate of radiative recombination at matter temperature T_e and electron and proton density n_e may be written as

$$-\frac{dn_e}{dt} = \alpha(T_e)n_e^{\,2},$$

from which a characteristic recombination time is

$$t_r = (\alpha n)^{-1} . \tag{10}$$

The recombination coefficient α is an average over the recombination cross-section multiplied by the velocity. In computing α recombinations direct to the ground state should be ignored because the recombination photon has short mean free path against ionizing another atom. A table of recombination coefficients is given by Boardman.[9]

The relevant number is the product of the recombination time (10) with the expansion rate $\dot{a}/a$, and it is interesting to evaluate it first for the present state of the Universe. This number is listed in Table VII-2, for

TABLE VII-2

PLASMA RECOMBINATION IN THE PRESENT UNIVERSE

Temperature T_e	Ht_r
100 K	$0.066\ (\Omega h)^{-1}$
1000	0.28
3000	0.56
10000	1.3

the present Universe, where the density parameter Ω (eq. 3) refers to the density of a uniformly distributed intergalactic plasma. It is apparent that if the plasma is reasonably hot or not too dense, $\Omega h < 1$, it need not recombine. On the other hand t_r at fixed temperature varies with redshift as $(1+z)^{-3}$ while the expansion rate in the general relativity cosmologies varies typically as $(1+z)^{3/2}$, so at moderately high redshift any appreciable amount of reasonably cool intergalactic plasma would be expected to recombine in the absence of ionizing radiation.

The rate of ionization of hydrogen atoms in blackbody radiation at temperature T may be written as

$$\frac{dn_H}{dt} = -\beta(T)\, n_H ,$$

where β is determined by the condition that at equilibrium, when the ionization is given by (9), the rates of ionization and recombination balance. This gives

$$\beta = \alpha(T)\,(2\pi\, m_e kT)^{3/2}\, e^{-B/kT}/h^3 \,. \tag{11}$$

Near the transition in the equilibrium from ionized to neutral matter this formula is not adequate because there are fewer thermal ionizing photons than nucleons, so if the matter were given as atomic hydrogen at that epoch it would not be ionized unless there were some relaxation process to keep the high energy tail of the Planck distribution filled. When $T < 10^5$ K the ratio of the number of ionizing blackbody photons to nucleons is, by equation (V-7) and (1),

$$\frac{n_\gamma}{n} \cong \frac{16\pi}{n}\left(\frac{kT}{hc}\right)^3 e^{-B/kT}$$

$$= \frac{3\times 10^7}{\Omega h^2}\, e^{-160000/T} \,.$$

This ratio is unity at $T = 10^4$ K, and increases rapidly with increasing T. Therefore when $T \gtrsim 10^4$ K the mean life of an atom against thermal photodissociation is given by β^{-1} (eq. 11) where α includes recombinations direct to the ground state. The mean life β^{-1} for $\Omega h^2 = 1$ is plotted as curve I in Figure VII-1. It joins onto the recombination time (eq. 10) plotted as curve R in the figure. For comparison the cosmological model age is plotted as a function of time for a cosmologically flat model with $h = 1$. This closely approximates what we have called the Einstein-deSitter model, except that it contains blackbody radiation at $T_0 = 2.7$ K, so at high redshift the model is radiation-dominated, causing the bend in the curve at $z \sim 10^4$. For other general relativity models with $\Lambda = 0$ the curves would be very similar. Evidently, if the expansion rate is not wildly different from the general relativity prediction the matter ought to be ionized at redshift $z \gtrsim 4000$.

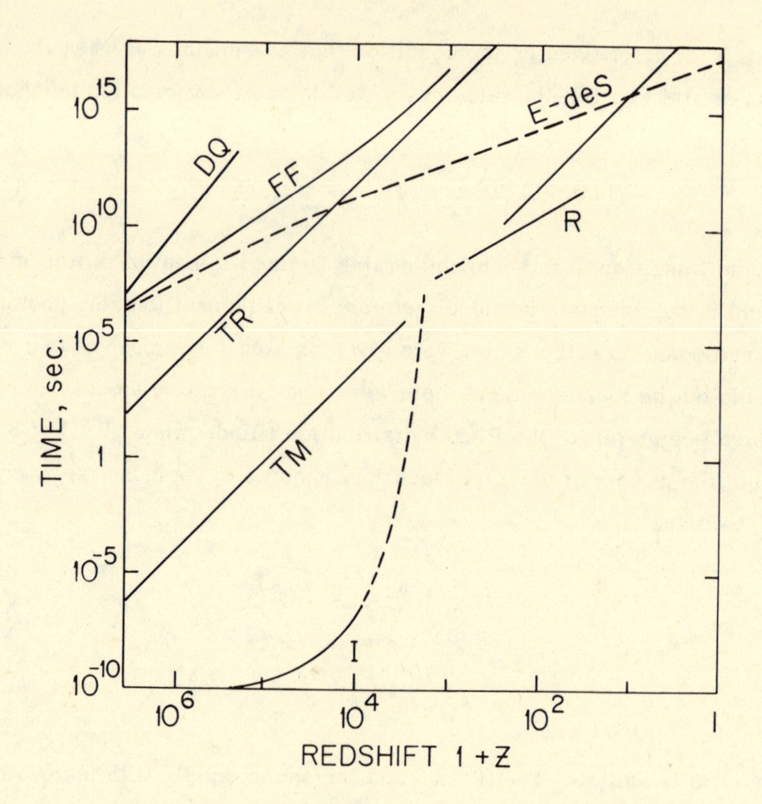

Fig. VII-1. Relaxation Times for the Cosmological Model. The solid lines give characteristic relaxation times for processes described in the text. The upper broken line is the expansion time in a Lemaître cosmological model.

iii) *Matter Temperature and Thomson Drag*

At $z \sim 1000$ the dominant process tending to keep the matter temperature equal to the radiation temperature is Thomson scattering by the free electrons. The heat exchange between matter and radiation by this process comes about as follows. Consider an electron with peculiar velocity $v << c$. In the electron rest frame the radiation temperature looks like (eq. V-39)

$$T(\theta) = T(1 + \frac{v}{c} \cos \theta) .$$

The radiation brightness (energy flux per steradian) is

$$i(\theta) = aT(\theta)^4 \, c/4\pi ,$$

and the momentum flux per steradian is i/c. The Thomson scattering cross-section is

$$\sigma_T = 6.65 \times 10^{-25} \text{ cm}^2 ,$$

to the desired accuracy $(v << c)$ independent of frequency. Since in the electron frame there is no front-back asymmetry in the scattered radiation we have from these equations that the rate of transfer of momentum to the electron is

$$\begin{aligned} F &= -\int \sigma_T \frac{i(\theta)}{c} \cos\theta \, d\Omega \\ &= -\frac{4}{3} \sigma_T \, aT^4 \, \frac{v}{c} . \end{aligned} \tag{12}$$

This is the radiation drag force on the electron.

It will be noted that in this process there is no correction for stimulated emission. To see this note that the rate for scattering a photon from the ith radiation mode to the jth mode is proportional to

$$\mathfrak{N}_i \, (1 + \mathfrak{N}_f) ,$$

where $\mathfrak{N}$ is the number of photons per mode (eq. IV-51). The first factor says that the rate is proportional to the number of photons, $\mathfrak{N}_i$, available to be scattered. The second factor is the sum of the spontaneous rate and the stimulated rate $(\propto \mathfrak{N}_f)$ to f. The opposite process, $f \to i$, contributes momentum transfer with the opposite sign, but the same rate coefficient, so to get the net momentum transfer we will be invited to sum over terms proportional to

$$\mathfrak{N}_i(1 + \mathfrak{N}_f) - \mathfrak{N}_f(1 + \mathfrak{N}_i) = \mathfrak{N}_i - \mathfrak{N}_f .$$

But the right hand side is just what one would have written down ignoring stimulated emission.

The mean rate of loss of energy per electron due to radiation drag, averaged over the electron velocity distribution, is by (12)

$$<-Fv> = \frac{4}{3}\sigma_T\, aT^4 <v^2>/c$$
$$= 4\,\sigma_T\, aT^4 k\, T_e/m_e c, \tag{13}$$

where the electron temperature has been defined by the usual formula

$$\frac{1}{2} m_e <v^2> = \frac{3}{2} kT_e .$$

At the same time electrons are gaining energy through the Brownian fluctuation in the radiation pressure on them, and when $T = T_e$ evidently the rate of loss (13) must equal the rate of gain, so the net rate of transfer of energy from radiation to matter per electron has to be[10]

$$\frac{dQ}{dt} = 4\,\sigma_T\, aT^4 k\,(T - T_e)/m_e c . \tag{14}$$

The electrons transfer energy to the ions by Coulomb scattering. The energy transfer rate is derived by Spitzer.[11] One finds that if the temperature is not too high, $T \lesssim 10^6$ K, the rate of energy transfer among electrons, ions and atoms is greater than (14), so it can be assumed that the matter is at uniform temperature T_e. If the matter is fully ionized, the matter kinetic energy is $3\,kT_e$ per electron, and (14) gives

$$\dot{T}_e = \frac{4\,\sigma_T\, aT^4}{3\, m_e c}(T - T_e). \tag{15}$$

The characteristic time for relaxation of T_e to T is therefore

$$t_T = 3\, m_e c/(4\,\sigma_T\, aT^4).$$

This is plotted as line TM in Figure VII-1 (where it is assumed that the matter is fully ionized).

iv) *Thomson Scattering and the Radiation Spectrum*

We have assumed so far that the radiation has a thermal spectrum. If it does not an important process by which the spectrum can evolve is

Thomson scattering, the same process considered in the last section, but here with attention directed to the energy change of the photon in the scattering. This has been analyzed by Kompaneets[12] and Weymann.[10]

The goal of the calculation simply is to compute the probability distribution function of the energy of a scattered photon for a given incident photon scattered by an electron moving with velocity $v << c$. The computation is somewhat lengthy because we have to worry about stimulated emission effects. It is simplest to compute in the electron rest frame. The photon distribution function $\mathfrak{N}$ is a Lorentz invariant, as is directly apparent from the fact that blackbody radiation viewed in a moving frame looks blackbody (eq. V-39), so the Lorentz transformations are simple.

Consider those free electrons moving in a given direction within a small solid angle, and with speed within a small range of some central value v. The number density of these electrons is δn. We will fix coordinates with positive X axis along the electron velocity. Then the Lorentz transformation laws for coordinates, photon frequency ν, and photon propagation angle θ relative to the X axis are

$$t = (t_1 + vX_1/c^2)/(1 - v^2/c^2)^{1/2}, \tag{16}$$

$$\nu = \nu_1 (1 + v/c \cos\theta_1)/(1 - v^2/c^2)^{1/2}, \tag{17}$$

$$\cos\theta_1 = (\cos\theta - v/c)[1 - v/c \cos\theta]^{-1}, \tag{18}$$

where the subscript 1 refers to the electron rest frame. The goal will be to compute to the same order as (12), that is, to second order in v/c and first order in $h\nu/m_e c^2$.

The invariance of $\mathfrak{N}$ says

$$\mathfrak{N}_1(\nu_1, \theta_1, t_1, X_1) = \mathfrak{N}(\nu, t = (t_1 + vX_1/c^2)/(1 - v^2/c^2)^{1/2}). \tag{19}$$

In the original comoving coordinate frame $\mathfrak{N}$ is supposed to be homogeneous and isotropic so we only have to worry about the arguments ν and t. We have from this equation that the scattering rates observed in electron rest frame and comoving coordinate frame are related by

$$\frac{d\mathcal{N}_1}{dt_1} = \frac{\partial \mathcal{N}_1}{\partial t_1} + c \cdot \nabla_1 \mathcal{N}_1$$
$$= \frac{\partial \mathcal{N}}{\partial t}\left(1 + \frac{v}{c}\cos\theta\right)\left(1 - \frac{v^2}{c^2}\right)^{-1/2} . \tag{20}$$

The scattering rate determined by an observer at rest in the electron frame is

$$\frac{d}{dt}\mathcal{N}_1(\theta_1, \nu_1) = c\,\delta n_1 \int \frac{d\sigma}{d\Omega_1'} d\Omega_1'$$
$$[(1+\mathcal{N}_1(\theta_1,\nu_1))\mathcal{N}_1(\theta_1',\nu_1+\delta\nu_1)\left(\frac{\nu_1+\delta\nu_1}{\nu_1}\right)^2 \frac{\partial}{\partial\nu_1}(\nu_1+\delta\nu_1) \tag{21}$$
$$- (1+\mathcal{N}_1(\theta_1',\nu_1-\delta\nu_1))\,\mathcal{N}_1(\theta_1,\nu_1)] .$$

Here $\delta\nu_1$ is the Compton expression for the energy transfer from radiation to electron when the electron initially is at rest,

$$\delta\nu_1 = \frac{h\nu_1{}^2}{m_e c^2}(1 - \cos\Theta) , \tag{22}$$

where Θ is the photon scattering angle, and we are keeping terms of first order in $h\nu_1/m_e c^2$. The first term in the square brackets in (21) is the rate at which photons are scattered from incident angle θ_1' into the beam. The first factor in this first term is the correction factor for stimulated emission. The second factor is the number of photons available to be scattered. The last two factors correct this number for the greater phase space at the higher frequency $\nu_1 + \delta\nu_1$. Similarly the second term represents the rate of scattering out of the beam.

To simplify (21) the trick is to expand the $\mathcal{N}$'s in a power series in the small frequency shift $\delta\nu_1$. First, however, it is convenient to make a change in notation. By (17) and (19) we can write

$$\mathcal{N}_1(\theta_1, \nu_1) = \mathcal{N}\left(\nu_1\left(1 + \frac{v}{c}\cos\theta_1\right)\left(1 - \frac{v^2}{c^2}\right)^{-1/2}\right)$$

$$\equiv \mathcal{N}(\nu).$$

For the distribution function at the other angle θ_1' we can write

$$\mathcal{N}_1(\theta_1', \nu_1) = \mathcal{N}\left(\nu_1\left(1 + \frac{v}{c}\cos\theta_1'\right)\left(1 - \frac{v^2}{c^2}\right)^{-1/2}\right)$$

$$\equiv \mathcal{N}\left(\nu\left(1 + \frac{v}{c}\cos\theta_1'\right)\left(1 + \frac{v}{c}\cos\theta_1\right)^{-1}\right),$$

and then we can expand this as a power series in v/c,

$$\mathcal{N}_1(\theta_1', \nu_1) = \mathcal{N}(\nu) + \frac{\partial\mathcal{N}}{\partial\nu}\nu\left[\frac{v}{c}(\cos\theta_1' - \cos\theta_1) + \frac{v^2}{c^2}\cos\theta_1(\cos\theta_1 - \cos\theta_1')\right]$$

$$+ \frac{1}{2}\frac{\partial^2\mathcal{N}}{\partial\nu^2}\nu^2\frac{v^2}{c^2}(\cos\theta_1' - \cos\theta_1)^2, \tag{23}$$

valid to the desired order v^2/c^2. When these expressions are substituted into the terms in the square parentheses in (21), and the expression expanded out to second order in v/c and first order in $\delta\nu/\nu$ (which means dropping terms like $(v/c)(\delta\nu/\nu)$), the result is

$$\begin{aligned}[\,] = \nu\frac{\partial\mathcal{N}}{\partial\nu}&\left[\frac{v}{c}(\cos\theta_1' - \cos\theta_1) + \frac{v^2}{c^2}\cos\theta_1(\cos\theta_1 - \cos\theta_1')\right] \\ &+ \frac{1}{2}\frac{\partial^2\mathcal{N}}{\partial\nu^2}\nu^2\frac{v^2}{c^2}(\cos\theta_1' - \cos\theta_1)^2 \\ &+ \left[4\mathcal{N}(\nu)(1 + \mathcal{N}(\nu)) + (1 + 2\mathcal{N}(\nu))\frac{\partial\mathcal{N}(\nu)}{\partial\nu}\nu\right]\frac{h\nu}{m_e c^2}(1 - \cos\Theta).\end{aligned} \tag{24}$$

Because this expression is now of first order in v/c we can use the ordinary Thomson differential scattering cross-section in (21), and it is readily seen therefore that all the terms in (24) linear in $\cos\theta_1'$ or $\cos\Theta$ vanish in the integral over $d\Omega_1'$. Next, we must use (20) to get $\frac{\partial\mathcal{N}}{\partial t}$ and then average over the incident photon direction θ. Because this integrand is of order v/c we can write (18) as

$$\cos\theta = \cos\theta_1 - \frac{v}{c}\cos^2\theta_1 + \frac{v}{c},$$

and use, for the average over θ,

$$d\cos\theta = (1 - 2\frac{v}{c}\cos\theta_1)\, d\cos\theta_1 .$$

Taking account of the factor $1 - v/c \cos\theta$ (= $1 - v/c \cos\theta_1$ to the desired accuracy) from (20) we find that the first term in the right hand side of (24) averaged over θ and integrated over scattering angle gives

$$(c\,\delta n\,\sigma_T)\,\frac{4}{3}\,\frac{v^2}{c^2}\,\nu\,\frac{\partial \mathfrak{N}}{\partial \nu} .$$

For the second term in (24) the average over θ for fixed scattering angle just reduces $\cos^2\theta_1$ and $\cos^2\theta_1'$ to 1/3 each. After averaging over the electron velocity we find that the end result, valid to order $(v/c)^2$ and $h\nu/m_ec^2$, is

$$\frac{\partial \mathfrak{N}}{\partial t}(\nu, t) = \sigma_T n_e c \left\{ \left[4\mathfrak{N}(1+\mathfrak{N}) + (1+2\mathfrak{N})\frac{\partial \mathfrak{N}}{\partial \nu}\nu \right] \frac{h\nu}{m_e c^2} + \frac{\langle v^2 \rangle}{c^2}\left[\frac{1}{3}\nu^2 \frac{\partial^2 \mathfrak{N}}{\partial \nu^2} + \frac{4}{3}\nu\frac{\partial \mathfrak{N}}{\partial \nu} \right] \right\}. \tag{25}$$

This is the desired equation for the time rate of change of the radiation distribution function due to scattering by free electrons with mean square velocity $\langle v^2 \rangle$.[10,12]

A comforting check of (25) is the time independent solution. If we define as before

$$\frac{1}{2} m_e \langle v^2 \rangle = \frac{3}{2} kT_e , \tag{26}$$

we find that $\partial \mathfrak{N}/\partial t = 0$ if

$$\mathfrak{N} = (e^{(h\nu - \mu)/kT_e} - 1)^{-1} , \tag{27}$$

where μ is a constant. If $\mu = 0$ this describes blackbody radiation in thermal equilibrium with the matter. If $\mu < 0$ (27) is a Bose distribution with non-zero chemical potential, corresponding to statistical equilibrium with a conserved particle number.[10] This is as it should be because the non-relativistic scattering process considered here conserves photons.

It will be noted that the expansion of the Universe preserves the functional form of (27). Because $\nu \propto a(t)^{-1}$, we have $T \propto a(t)^{-1}$ as before, and $\mu \propto a(t)^{-1}$.

Because photons are conserved the Compton scattering process cannot produce blackbody radiation from some initially given radiation spectrum if there are too few photons present for the amount of energy. This corresponds to the case $\mu < 0$, and it yields a distribution function $i \propto \nu^3$ at low frequency, $h\nu < (\mu, kT)$. If there are too many photons for the energy μ cannot go above the lowest energy state because this would make (27) singular, so the relaxation drives the excess photons down to very low energy (Bose-Einstein condensation), and the resulting distribution looks blackbody.

The net rate of transfer of energy from matter to radiation per unit volume is found by multiplying (25) by $h\nu$ and integrating over phase space. The result after integration by parts is

$$\begin{aligned} - n_e \frac{dQ}{dt} &= \int h\nu \frac{\partial \mathfrak{N}}{\partial t} \cdot \frac{8\pi\, \nu^2\, d\nu}{c^3} \\ &= \sigma_T\, n_e c \left[\frac{4}{3} \frac{\langle v^2 \rangle}{c^2} U - \int \frac{8\pi\, \nu^2\, d\nu}{c^3} \frac{(h\nu)^2}{m_e c^2} \mathfrak{N}(1+\mathfrak{N}) \right], \end{aligned} \tag{28}$$

where

$$U = \int \frac{8\pi\, h\nu^3}{c^3} \mathfrak{N}\, d\nu$$

is the radiation energy density. The first term in (28) agrees with (13). The generalization of (15) is

$$\frac{d\,T_e}{dt} = \frac{4\,\sigma_T\, U}{3\, m_e c} \left[\int \frac{8\pi\, \nu^2\, d\nu}{c^3} \frac{(h\nu)^2}{4\, kU} \mathfrak{N}(1+\mathfrak{N}) - T_e \right], \tag{29}$$

giving the relaxation time

$$t_r = 3\, m_e c/4\, \sigma_T\, U\, .$$

If the radiation energy has not been appreciably augmented by processes like matter-antimatter annihilation then $U \propto (1+z)^4$, and curve TM of Figure VII-1 applies to the more general equation (29).

By equation (25) a measure of the time for the radiation to relax to the equilibrium spectrum (27) is

$$t_c = m_e c/(\sigma_T\, n_e\, kT)\, . \tag{30}$$

This is plotted as line TR in Figure VII-1 for the case $\Omega h^2 = 1$.

Solutions of (25) have been discussed by Kompaneets,[12] Weymann,[2] and Zeldovich and Sunyaev.[13] An interesting and simple case is that the radiation spectrum is initially given as the greybody form

$$\mathcal{N} = \frac{(T/T_g)}{e^{h\nu/kT_g} - 1}\, . \tag{31}$$

This would happen if the radiation initially were thermal but concentrated in small hot spots occupying a fraction T/T_g of space, so that when the radiation spreads over all space it is diluted by the factor T/T_g. The low frequency part of (31) looks like a Rayleigh-Jeans spectrum at temperature T, in agreement with observation of the microwave background, but the high frequency cutoff is higher than the thermal value (eq. V-10) by the factor T_g/T. On substituting (31) into (29) and setting the right hand side equal to zero we find that the equilibrium matter temperature under Thomson scattering is

$$T_e = T_g - \tfrac{1}{4}(T_g - T)\left[\int X^4\, dX/(e^X - 1)^2\right]\left[\int X^3\, dX/(e^X - 1)\right]^{-1}$$

$$= 0.958\, T_g + 0.042\, T\, .$$

According to Figure VII-1 if the expansion rate is given at least roughly by the Einstein-deSitter model the matter temperature should relax to this equilibrium value when $z > 1000$.

To see how (31) evolves under Thomson scattering consider the limiting case $T_g >> (T, h\nu)$. This gives

$$\mathfrak{N} = kT/h\nu ,$$

$$\frac{\partial \mathfrak{N}}{\partial t} = -2\,\sigma_T\, n_e c \left(\frac{kT_e}{m_e c^2}\right)\left(\frac{kT}{h\nu}\right).$$

That is, the long wavelength tail is decreasing as low frequency photons move up to fill the peak of the distribution (27), but at first the tail preserves its Rayleigh-Jeans spectrum, the Rayleigh-Jeans temperature cooling according to (cf. eq. 30)

$$\dot{T}/T = -2\,\sigma_T\, n_e c \left(\frac{kT_e}{m_e c^2}\right). \tag{32}$$

v) *Photon Production*

If in the early Universe there were too few photons for the amount of energy present the radiation distribution function could relax to blackbody only if photons were produced rapidly enough, and the most rapid processes would seem to be free-free bremsstrahlung emission. A near competitor near $T = 3000$ K is bound-free emission,

$$e + p \rightarrow H + \gamma ,$$

where if the recombination is to an excited state and if the atom can decay to the ground state before it is ionized again the gain is one photon. If ionization from the excited state could be ignored the photon production rate would be comparable to that from free-free emission, but since the ionization rate is greater than $10^8\ \mathrm{sec}^{-1}$ when $T \gtrsim 10^4$ K, the net production in fact is much smaller. We will concentrate therefore on free-free emission.

A measure of the time required to produce a thermal photon abundance is

$$t_{ff} = \left[60.2\left(\frac{kT}{hc}\right)^3\right]\left[\int 4\pi\, j_\nu\, d\nu/h\nu\right]^{-1},$$

where the first factor is the thermal photon number density at temperature T (eq. V-7). Because j_ν (eq. IV-66) is nearly constant longward of hc/kT there is an approximately logarithmic divergence at long wavelength, but this does not mean that there are unlimited numbers of photons available because at long wavelength the plasma is optically thick, so photons are absorbed almost as fast as they are created. The cut-off due to self-absorption is roughly at the frequency ν_c where the mean free time for free-free absorption is equal to the characteristic time defined by (32),

$$\kappa(\nu_c)\, c = 2\, \sigma_T\, n_e\, kT/m_e c\,. \qquad (33)$$

Now there are two effects. In the band of frequencies ν_c to kT/h photons produced by bremsstrahlung emission almost all survive and add to the store. At the same time the radiation longward of the cut-off is moving to higher frequency according to equation (32). One finds that the latter effect is comparable to but somewhat smaller than the former, which may be seen as follows. On writing the free-free luminosity as

$$j_\nu = A\, e^{-h\nu/kT},$$

we have that the rate of production of photons at $\nu > \nu_c$ is

$$\frac{d}{dt} n(>\nu_c) = \int 4\pi\, j_\nu/h\nu\, d\nu$$

$$\cong \frac{4\pi A}{h} \ln \frac{kT}{h\nu_c}\,.$$

At $\nu \sim \nu_c$ the radiation spectrum has relaxed to thermal equilibrium, photon number density

$$n_\nu = 8\pi\, kT\, \nu/(hc^3)\,.$$

These photons are moving to higher frequency at the rate (eq. 32)

$$\dot{\nu}/\nu = \sigma_T n_e c \left(\frac{kT}{m_e c^2}\right),$$

so the rate at which this process supplies photons to $\nu > \nu_c$ is

$$\frac{d}{dt} n(>\nu_c) = n(\nu_c)\,\dot{\nu}(\nu_c)$$

$$= \frac{8\pi\, kT}{hc^3} \nu_c^{\,2}\, \sigma_T n_e c \left(\frac{kT}{m_e c^2}\right).$$

By (33) we can rewrite this as

$$\frac{d}{dt} n(>\nu_c) = \frac{4\pi\, kT}{hc^2} \nu_c^{\,2}\, \kappa(\nu_c),$$

but since by detailed balance

$$j = \frac{2kT\,\nu^2}{c^2}\,\kappa$$

this reduces to

$$\frac{d}{dt} n(>\nu_c) = 2\pi A/h,$$

which differs from the direct photon production by the factor $\ln\,(kT/h\,\nu_c)^2$.

By equations (33) and (IV-65) the cut-off is

$$\frac{h\nu_c}{kT} \sim \frac{300(\Omega h^2)^{1/2}}{T^{3/4}},$$

where T is in degrees Kelvin, and the photon production time is

$$t_{ff} \sim \frac{6.0 \times 10^{24}\ (\Omega h^2)^{-2}}{T^{5/2}\ \ln\,[T^{3/4}/(300(\Omega h^2)^{1/2})]}\,\text{sec.}$$

This is plotted as line FF in Figure VII-1 on the assumption $\Omega h^2 = 1$. The figure shows that at redshift $z \sim 10^6$ the photon production rate is just about enough to assure relaxation to a blackbody spectrum if $\Omega h^2 \sim 1$.

In a low density model, $\Omega h^2 \sim 10^{-2}$, apparently photon production would depend on the next fastest process, double quantum emission,

$$\gamma + e \rightarrow \gamma + \gamma + e\,,$$

which becomes important when the electrons start to get relativistic. The cross-section for double quantum emission given by Heitler is, in order of magnitude,

$$\sigma_D = \frac{\sigma_T}{137}\left(\frac{kT}{m_e c^2}\right)^2 ,$$

which gives the characteristic time

$$n_\gamma/\dot{n}_\gamma = (\sigma_D\, n_e c)^{-1}$$

$$\sim 10^{42}\, T^{-5}\, (\Omega h^2)^{-1} \text{ sec.}$$

This number is not very precise but because the temperature dependence is so steep it hardly matters. It is plotted as curve DQ in Figure VII-1.

vi) *Discussion*

One might imagine that the early Universe only settled down to a reasonable approximation to a homogeneous and isotropic world model at some epoch labelled by redshift z^*. Then if we can estimate the subsequent expansion rate we can use Figure VII-1 to see how large z^* would have to be to assure a reasonable approximation to a thermal Fireball background now. If the expansion rate is given by the general relativity model, and if $z^* \sim 10^4$, then the matter ought to be ionized and Thomson scattering by the free electrons should relax the radiation distribution function. If at z^* there are too many photons for the energy this equilibrium is a blackbody distribution, the excess photons being driven down to low frequency. If at z^* there are too few photons we must go to $z^* \sim 10^6$ if the Universe has high density, $z^* \sim 10^7$ if low density, to permit production of the extra photons.

Further complications are considered in the following sections.

b) *Dynamic History of Matter in the Big Bang Model*

The Lemaître cosmological model assumes a perfectly homogeneous mass distribution, which of course is manifestly untrue over a distance scale as large as 10 Mpc. This defect can be corrected in a reasonable approximation, and what one finds is that if irregularities are present (as they are), they tend to grow worse as time goes on. In effect the Universe is gravitationally unstable. In a sense the miracle is not that the Universe contains irregularities like galaxies, but that the Universe is as approximately uniform and regular as it is. As will be described the miracle is to some extent relieved by the effect of the Primeval Fireball, which can turn off the instability over an appreciable span of time in the early Universe. An interesting variant of the density irregularity problem is the speculation that the matter in the early Universe was turbulent, and that the galaxies may even be the fossil remnants of turbulent eddies. The purpose of this section is to discuss some of these dynamic effects.

i) *Small-Scale Irregularities*

In all the following discussion it will be assumed that the dimensions of density irregularities are much less than ct, t being the cosmic time in the model, and it will be assumed that the motion of the matter is nonrelativistic, peculiar velocity v everywhere much less than c. This would appear to be a reasonable assumption at least in the present Universe, and it has the great virtue that all the complications of general relativity theory may be ignored, which may be seen as follows. As in Chapter VI we will simplify the notation here by choosing units such that the velocity of light is unity.

Let us start with a freely moving observer more or less at rest relative to the matter about him. We will set up a coordinate system with this observer at the spatial origin, $x^{\alpha} = 0$. Then coordinates may be chosen so that, at the origin, the components of the metric tensor g_{ij} reduce to their ordinary flat space Minkowski form η_{ij}, and the gradient of g_{ij} vanishes. This is the ordinary locally Minkowski coordinate system the observer might set up in his neighborhood. The coordinate system warps due to curvature with increasing distance from the observer, so the metric tensor looks like

$$g_{ij} = \eta_{ij} + h_{ij}\,(x, t),$$

where h_{ij} is of second order in the distance x from the origin. If x is kept sufficiently small h_{ij} can then be regarded as a small perturbation, determined by the weak field limit of the gravitational field equations. The great simplification is that when $v << c$ the weak field limit is just Newtonian mechanics. For example, in this limit the geodesic equations of motion (eq. VI-2) reduce to

$$\frac{d^2 x^\alpha}{dt^2} = -\frac{1}{2}\, h_{00,\,\alpha}\,. \tag{34}$$

In the weak field approximation the zero-zero component of R_{ij} is[14]

$$R_{00} = \frac{1}{2}\,\eta^{ij}\,(h_{00,\,ij} + h_{ij,\,00} - 2h_{0i,\,0j})\,.$$

On evaluating this expression at the origin, and recalling that h_{ij} is second order in x^α, we find that the zero-zero part of Einstein's gravitational field equations is

$$\nabla^2\left(\frac{1}{2}\, h_{00}\right) = 4\pi\, G(\rho + 3P)\,. \tag{35}$$

This equation fixes h_{00} (up to a homogeneous solution) to the required accuracy, second order in x^α. It is apparent from (34) and (35) that the usual Newtonian gravitational potential is

$$\phi = h_{00}/2\,. \tag{36}$$

Equations (34) and (35) are valid when the distances and velocities are small. When these equations are applied to the expanding Universe evidently x must be much less that the Hubble length H^{-1}, for otherwise recession velocities would become comparable to the velocity of light. The solution to (35) is, in order of magnitude,

$$h_{00} \sim G\rho\, x^2 \sim H^2\, x^2\,,$$

where the second equation follows because we know from observation that H^2 is roughly the same order of magnitude as $G\rho$. Thus when $x << H^{-1}$, $h_{00} << 1$, so the weak field approximation applies. This is a generous condition. In the present epoch $H^{-1} \sim 3000$ Mpc, while the size of the largest well-established structure we want to study is ~ 10 Mpc (Local Supercluster).

ii) *Zero Pressure Linear Density Perturbations*

A simple but important problem is the time variation of small density perturbations, $\delta\rho/\rho << 1$, when pressure may be neglected. The matter density and velocity in the locally Minkowski coordinates defined above may be written as

$$\begin{aligned} \rho_1 (\mathbf{x}, t) &= \rho(t) [1 + \delta(\mathbf{x}, t)], \\ \mathbf{u} &= \frac{\dot{a}}{a} \mathbf{x} + \mathbf{v} (\mathbf{x}, t), \end{aligned} \tag{37}$$

where $a(t)$ and $\rho(t)$ are the expansion parameter and matter density in the unperturbed model. We will compute to first order in the perturbations δ and $\mathbf{v}$.

The equation of conservation of mass is

$$\frac{\partial \rho_1}{\partial t} + \nabla_{\mathbf{x}} \cdot (\rho_1 \mathbf{u}) = 0 .$$

When equations (37) are substituted into this equation we obtain the usual unperturbed expression

$$\frac{d\rho(t)}{dt} + 3 \frac{\dot{a}}{a} \rho(t) = 0 ,$$

and the first order part

$$\left.\frac{\partial \delta}{\partial t}\right)_{\mathbf{x}} + \frac{\dot{a}}{a} \mathbf{x} \cdot \nabla_{\mathbf{x}} \delta + \nabla_{\mathbf{x}} \cdot \mathbf{v} = 0 . \tag{38}$$

Now it is convenient to change spatial variables from the locally Minkowski coordinates x^α to coordinates r^α comoving in the unperturbed model, defined by the equation

$$x^\alpha \equiv a(t)\, r^\alpha . \tag{39}$$

Then according to the usual rules of differentiation,

$$\left.\frac{\partial}{\partial t}\right)_r = \left.\frac{\partial}{\partial t}\right)_x + \frac{\dot a}{a} x^\alpha \left.\frac{\partial}{\partial x^\alpha}\right)_t ,$$

so (38) becomes

$$\left.\frac{\partial \delta}{\partial t}\right)_r + \frac{\nabla \cdot v}{a} = 0 , \tag{40}$$

where the gradient sign without the subscript means derivative with respect to r^α.

The equations of motion of the matter are

$$\frac{\partial \mathbf{u}}{\partial t} + (\mathbf{u} \cdot \nabla)\, \mathbf{u} = -\nabla_x \phi , \tag{41}$$

where the Newtonian potential ϕ (eqs. 35, 36) may be written as the sum of an unperturbed part and a perturbed part,

$$\begin{aligned} \phi &= \frac{2\pi}{3} G\rho\, x^2 + \chi , \\ \nabla_x{}^2 \chi &= 4\pi\, G\rho\delta . \end{aligned} \tag{42}$$

When (37) is substituted into (41) and use made of (42) we find that the unperturbed part of the equation is the usual expression for the acceleration of the expansion parameter a(t) (eq. I-11), and the perturbed part is

$$\left.\frac{\partial v}{\partial t}\right)_x + \frac{\dot a}{a}(x \cdot \nabla_x) v + \frac{\dot a}{a} v = -\nabla_x \chi ,$$

which with the change of coordinates from x to r becomes

$$\frac{\partial v}{\partial t} + \frac{\dot a}{a} v = -\frac{\nabla \chi}{a} .$$

The divergence of this equation with (40) and (42) yields

$$\frac{\partial^2\delta}{\partial t^2} + 2\frac{\dot{a}}{a}\frac{\partial\delta}{\partial t} = 4\pi\, G\rho(t)\,\delta\,. \tag{43}$$

When the expansion rate of the Universe approximates the Einstein-deSitter model, which should be valid for $z \gtrsim 30$ if $\Omega \gtrsim 0.03$, the density is $\rho = 3(\dot{a}/a)^2/8\pi\, G$, $\dot{a}/a = 2/3t$, and (43) reduces to

$$\frac{\partial^2\delta}{\partial t^2} + \frac{4}{3t}\frac{\partial\delta}{\partial t} = \frac{2\delta}{3t^2}\,. \tag{44}$$

The solution is

$$\delta = A(r)\, t^{2/3} + B(r)\, t^{-1}. \tag{45}$$

Equation (45) was first derived by Lifshitz in a classic general study of linear perturbations to a Lemaître cosmological model, and was first derived from the Newtonian approximation by Bonnor.[15] Lifshitz argued that the galaxies could not have formed from a gravitational instability because (45) grows too slowly, and his conclusion was endorsed by Bonnor and a number of other authors. Whatever the merit of this conclusion it had the unfortunate effect of obscuring a key point. The Universe now evidently is irregular, and to the extent that (45) is a good approximation we know the irregularity can only grow worse as time goes on. Turning this around, how could the Universe be as approximately regular as it is now in the large scale? Some possibilities are (a) the very early Universe, $t << t_0$, was smooth and uniform to miraculous precision, (b) there are stabilizing effects to counterbalance gravitation, (c) matter was injected into the Universe at a relatively recent epoch, so it hasn't had time to get completely messed up.

Some important stabilizing effects are described in the next two sections.

iii) *The Jeans Length*

The first important correction to (45) is the kinetic matter pressure. This adds to the right hand side of (41) the pressure force per unit volume,

$-(\nabla P)/\rho$. To find the size of the effect let us suppose the perturbation is isothermal, so by the ideal gas law

$$(\nabla P)/\rho = \frac{kT}{m_p} \nabla \delta ,$$

with m_p the mass of a hydrogen atom. With this pressure term (43) becomes

$$\frac{\partial^2 \delta}{\partial t^2} + 2 \frac{\dot{a}}{a} \frac{\partial \delta}{\partial t} = 4\pi G\rho \delta + \frac{kT}{m_p} \frac{\nabla^2 \delta}{a^2}. \tag{46}$$

The convenient way to understand the competition between the two terms on the right hand side of this equation is to represent the density perturbation as the Fourier series,

$$\delta = \sum \delta_k(t) e^{i\mathbf{k}\cdot\mathbf{r}} ,$$

where the proper wavelength belonging to wave number k is (eq. 39)

$$\lambda = 2\pi a/k . \tag{47}$$

Because the coefficients in (46) are independent of position the differential equations for the amplitudes separate, to give

$$\frac{d^2 \delta_k}{dt^2} + 2\frac{\dot{a}}{a} \frac{d\delta_k}{dt} = \left(4\pi G\rho - \frac{4\pi^2 kT}{m_p \lambda^2} \right) \delta_k . \tag{48}$$

When the wavelength is very long the new term in (48) is unimportant and we get back to (43). When λ is small it dominates the gravity term and δ oscillates like an acoustic wave. The wavelength separating these limiting cases is the critical Jeans length, where the right hand side of (48) vanishes,

$$\lambda_J \equiv \left(\frac{\pi kT}{G\rho\, m_p} \right)^{1/2} . \tag{49}$$

This critical length was first derived by Jeans, in 1902, using a method of approach like the one described here. As is notorious this approach is singular in Newtonian mechanics if one attempts to start from an unperturbed solution representing a homogeneous mass distribution, but it is pleasant to note that the Jeans criterion applies cleanly and directly in the Lemaître model.

The exciting application of (49) is that from "reasonable" values of ρ and T one can derive a length which one ought to be able to relate to some phenomenon, like galaxies. The problem is that different people have different ideas about what is reasonable. If one adopts $\rho = \rho_c = 2 \times 10^{29}\ h^2\ g\ cm^{-3}$, one finds

$$\lambda_J = 0.0046\ T^{1/2}\ h^{-1}\ \mathrm{Mpc}\ .$$

It has been pointed out that, if T were about 10^6 K, λ_J would be about 5 Mpc, interestingly close to the distance between big galaxies (eq. IV-3). Another suggestion for ρ and T is based on Figure VII-1. At $z \sim 1000$ matter should recombine, and, as described in the next section, should be set free to collect in gravitationally bound clouds. Earlier it is locked to the radiation. We think we know the matter temperature on decoupling, 3000 K, and the matter density, so we get another Jeans length, and a characteristic mass from

$$\mathfrak{M}_J = \rho \lambda_J{}^3\ .$$

The result is $\mathfrak{M}_J \sim 10^5 \mathfrak{M}_\odot$, well below the mass of an "ordinary" galaxy, but, it has been argued, interestingly close to the mass of a globular star cluster.[16]

The dynamic effect of a primeval magnetic field can be expressed in terms of an effective Jeans length in direct analogy with (49). A reasonable magnetic pressure would be the local value $B^2/8\pi$. (If the field were tangled the large-scale pressure would be one third of this value.) Taking account of the fact that the perturbation this time should be "adiabatic," $B^2 \propto \rho^{4/3}$, we find that the magnetic Jeans length is[17]

$$\lambda_B \sim (B^2/6\,G\rho^2)^{1/2} \sim 2.7\,B_{10}\,(\Omega h^2)^{-1}\,(1+z)^{-1}\ \mathrm{kpc},$$

where B_{10} is the present magnetic field measured in units of 10^{-10} Gauss. In writing down the redshift dependence it is assumed that the expansion conserved flux. It then follows that the characteristic mass $\rho\lambda_B{}^3$ is independent of epoch, and could of course be fitted to a "reasonable" mass with a not unreasonable B.

If the magnetic field has structure on a scale $< \lambda_B$ then the field can move the matter, so for example a tangled field would tend to straighten itself, pushing the matter into the pockets where the flux density initially was lowest. The interesting consequence would be that the parameter $B\rho^{-2/3}$ would be smallest in the Galaxies, where the matter is concentrated. Magnetic forces and the attendant magnetohydrodynamic phenomena could very well play an important role in the history of the Universe, but the unfortunate thing at the moment is that we do not know for sure whether there is or was a general intergalactic field, and if so how intense it might be (ref. IV-49).

iv) *Radiation Drag*

As long as the matter is neutral the only effect of the Primeval Fireball radiation is to increase somewhat the rate of expansion of the cosmological model. For example, in a radiation dominated model $(P = \rho c^2/3)$ where curvature may be neglected we have $\dot{a}/a = (2t)^{-1}$, and (43) yields $\delta \propto t^{\pm n}$, $n = (3/8)^{1/2}$, not very different from (45). However, if the matter is coupled to the radiation (because the radiation mean free path is short) it can very significantly alter the picture.[18]

At redshift $z \gtrsim 10000$ the Universe is radiation dominated, pressure comparable to ρc^2. Under this condition, and assuming matter and radiation are tightly coupled by scattering so they act like one fluid, the critical Jeans length (49) becomes $\lambda_J \sim ct$, much bigger than the scale of perturbation we are considering. That is, the radiation pressure cuts off the growth of non-relativistic perturbations.

The coupling between matter and radiation may be computed as follows. Consider a matter density perturbation with characteristic size much less than the Jeans length for matter and radiation considered as a single fluid. Then we can take the radiation to be uniformly distributed, and the matter density irregularity can grow only if the matter can slip by the radiation. If the matter is moving at velocity v through the radiation the radiation drag force per unit volume opposing this motion is (eq. 12)

$$\frac{4}{3}\frac{\sigma_T\, aT^4}{m_p c}\,\rho v\ ,$$

assuming the matter is fully ionized. When this force is added to (41), (46) becomes

$$\frac{\partial^2\delta}{\partial t^2} + \left(2\frac{\dot a}{a} + \frac{4}{3}\frac{\sigma_T\, aT^4}{m_p c}\right)\frac{\partial\delta}{\partial t} = 4\pi\, G\rho\,\delta + \frac{kT}{m_p}\frac{\nabla^2\delta}{a^2}\ . \qquad (50)$$

The ratio of the two terms in parentheses in the left hand side of this equation is

$$\frac{2\sigma_T\, aT^4}{3\, m_p(\dot a/a)\, c} = 1.1\times 10^{-6}\,(\Omega h^2)^{-1/2}\,(1+z)^{5/2}\ . \qquad (51)$$

For the expansion rate we have used $\dot a/a \cong H\Omega^{1/2}(1+z)^{3/2}$, which by I-10 is a good approximation for $100 \lesssim z \lesssim 10000$ because the curvature term may be neglected and $\rho \propto (1+z)^3$. When $z \gtrsim 1000$, (51) is large, $\gtrsim 30$, which means the radiation drag term dominates the left hand side of (50). If the pressure term on the right hand side also may be neglected ($\lambda >> \lambda_J$) then (50) reduces to the approximate form

$$\frac{4}{3}\frac{\sigma_T\, aT^4}{m_p c}\frac{\partial\delta}{\partial t} \cong 4\pi\, G\rho\delta\ ,$$

or

$$\delta(t) \cong \delta(0)\exp\int_0^t \frac{3\pi\, G\rho\, m_p c\, dt}{\sigma_T\, aT^4} \cong \exp\left(\frac{2.7\times 10^5\,(\Omega h^2)^{1/2}}{(1+z)^{5/2}}\right).$$

If, for example, $\Omega h^2 = 1$, this gives $\delta/\delta(0) = 1.0003$ at $T = 10000$ K, and $\delta/\delta(0) = 1.006$ at $T = 3000$ K. Up to this point the radiation drag cuts off the non-relativistic gravitational instability very effectively indeed.

When T falls to ~ 3000 K the matter should recombine, when this happens the coupling between matter and radiation is very much reduced, and matter density irregularities are set free to evolve according to equation (45) or (48). This is the basis for the primeval globular cluster picture mentioned in the previous section. If for some reason the matter does not recombine on schedule the radiation drag force is effective until (51) falls below unity, which happens at redshift

$$(1 + z) \sim 200\,(\Omega h^2)^{1/5}.$$

The Primeval Fireball, if indeed it exists, is of profound significance for the Big Bang picture because it stops non-relativistic irregularities from developing into bound systems until the Universe has expanded to modest density, $z \lesssim 1000$, $n \lesssim 10^3\ \mathrm{cm}^{-3}$. We can at least understand why non-relativistic objects like galaxies did not form out at the density of iron, say, for when the nucleon density was this high the matter was firmly locked to the radiation and the enormous radiation pressure dominated gravity. This is an important step forward but of course our troubles are by no means over. We have assumed all along that $\lambda << ct$. When this condition is not satisfied radiation pressure is ineffective and we come back to the question, what prevented the growth of irregularities of very large scale? The answer is not clear, and for that matter neither is the question.[19,20]

v) *Primeval Turbulence*

There are two very different lines of thought on what the early Universe might have been like in the Big Bang cosmology. One is that the Universe must have been very smooth and regular because it is unstable, if only mildly so, so the approximate homogeneity of the Universe now means that it used to be much more homogeneous.[19] The opposing line of thought is that the early Universe was chaotic.[20] The main point of attraction of this latter idea is that chaos requires no special initial values, as opposed to the very precise symmetry *ab initio* implied by the first argument.

A second reason for studying primeval turbulence is the thought that the galaxies may be the fossil remnants of primeval turbulence eddies. This idea has enjoyed wide and enduring attention. Turbulence was introduced to cosmology by von Weizsäcker,[21] and the idea was taken up by Gamow when he became convinced that linear density irregularities (eq. 45) grow too slowly to make galaxies.[22] Most recently the idea that turbulence may have played a role in galaxy formation has been discussed by Oort, Ozernoi and Chernin, Harrison, and Sato, Matsuda and Takeda.[23]

A full discussion of the primeval chaos concept apparently would be enormously difficult, and we are still far from it. It is easy to discuss a more restricted hypothesis, that in the early Universe the peculiar velocity of the matter, expressed as a deviation from a homogeneous isotropic model, is much less than the velocity of light. The very interesting property of "primeval turbulence" of this sort is that when viscous dissipation may be ignored the mean square turbulence velocity is constant, unaffected by the expansion of the Universe, as long as $P = \rho c^2/3$. This result was first obtained by Lifshitz in linear perturbation theory.[15] For characteristic "turbulent eddy" size $\lambda >> ct$ the linear approximation is entirely appropriate because the displacement of any fluid element is $vt << ct << \lambda$. When $\lambda << ct$ the motion is non-linear. The conservation of v^2 here is most easily understood from the requirement of angular momentum conservation. The angular momentum of a turbulent eddy with size λ, rotating with angular velocity ω, is $\sim \rho\lambda^5\omega$. But when $P = \rho c^2/3$, $\rho\lambda^4 \sim$ constant, so $\lambda\omega \sim v$ should be constant. A more explicit derivation, under the assumptions (1) $\lambda << ct$, (2) $v << c$, (3) $P = \rho c^2/3$, (4) ideal fluid, goes as follows.

Because the flow is subsonic the mass density is very nearly homogeneous (independent of position). It follows that the gravity source density (stress-energy tensor) is not sensibly different from what it would be in a homogeneous isotropic world model, because the source depends on ρ and P, which are not significantly irregular, and on v/c, which is much less than unity. Therefore we can use for the line element the homogeneous and isotropic form (VI-16). A final handy simplification is that we can ignore the curvature term $a(t)R$. To see this note that, by equation I-10, the radius of curvature in the present Universe $|a_0R/c|$ is greater than or comparable to $\left(\frac{8}{3}\pi G\rho\right)^{-1/2}$, for if it were very much smaller we would not have the observed rough agreement between H^2 and $8\pi G\rho/3$. Now as we go back in time $|a(t)R/c|$ decreases with decreasing expansion parameter a slower than $\left(\frac{8}{3}\pi G\rho\right)^{-1/2}$, so

$$\left|\frac{aR}{c}\right| >> \left(\frac{8}{3}\pi G\rho\right)^{-1/2} \cong a/\dot{a}, \; a << a_0 .$$

Since $a/\dot{a} \sim t$, this says $|aR| >> ct$. For the characteristic size λ of the turbulence eddy we want to consider, $\lambda << ct$, it follows that $\lambda << |aR|$. Thus over the region of space occupied by a turbulent eddy we have $r << R$, and (VI-16) is well approximated by

$$ds^2 = dt^2 - a(t)^2 \left[(dr^1)^2 + (dr^2)^2 + (dr^3)^2\right] . \tag{52}$$

Because the flow is incompressible it satisfies, in the coordinates of equation (52), the condition

$$\partial v^\alpha / \partial r^\alpha = 0 . \tag{53}$$

The equations of motion of the fluid are derived from the covariant constraint on the energy-momentum tensor,

$$0 = T_{i;j}^{\,j} = \frac{1}{(-g)^{1/2}} \frac{\partial}{\partial r^j} \left[T_i^{\,j} (-g)^{1/2}\right] - \frac{1}{2} g_{jk,i} T^{jk} . \tag{54}$$

The matter and radiation will be approximated as an ideal fluid,

$$T^{ij} = (\rho + P) u^i u^j - g^{ij} P , \tag{55}$$

where $u^0 \cong 1$ and the peculiar velocity observed at any point is related to u^α by

$$v^\alpha \equiv a \frac{dr^\alpha}{dt} = a u^\alpha . \tag{56}$$

When (55) is substituted into (54), with the metric tensor fixed by (52), and the $i = 0$ component of the resulting equation multiplied by v^α and subtracted from the $i = \alpha$ component, the result is

$$\left(\rho+\frac{P}{c^2}\right)\left[\frac{\partial v^\alpha}{\partial t}+\frac{v^\beta}{a}\frac{\partial v^\alpha}{\partial r^\beta}\right]+\left(\rho+\frac{P}{c^2}\right)\frac{\dot a}{a}v^\alpha+\frac{1}{a}\frac{\partial P}{\partial r^\alpha}+\frac{\partial P}{\partial t}\frac{v^\alpha}{c^2}=0. \qquad (57)$$

The expression in square parentheses in the first term is just the rate of change of the peculiar velocity of a chosen fluid element. It is multiplied by the inertial mass density, and the only non-obious thing here is that the pressure contributes to the inertial mass. The second term gives rise to the slowing of peculiar velocity (or momentum) due to expansion, equations (I-13, 15) and (VI-21). If $P = 0$ the equation says that the peculiar velocity $\propto a(t)^{-1}$, as before. The third term in (57) is the usual volume force due to the pressure gradient. The last term arises because we are supposed to evaluate the pressure gradient at fixed time not in the coordinates of (52) but in the rest frame of the fluid, and since P is varying with time there is a correction for the difference of time synchronization in the two coordinate systems.

We are assuming $P = \rho c^2/3$. Under this condition if (57) is multiplied by v^α and summed over α it yields

$$\frac{2}{3}\frac{\partial v^2}{\partial t}+\frac{2}{3}\frac{v^\beta}{a}\frac{\partial v^2}{\partial r^\beta}+\frac{4}{3}\frac{\dot a}{a}v^2+\frac{v^\alpha}{3a}\frac{\partial}{\partial r^\alpha}\ln\rho+\frac{v^2}{3}\frac{\partial \ln\rho}{\partial t}=0.$$

When this equation is averaged over space at fixed cosmic time t all the derivatives with respect to r^β vanish by integration by parts because of equation (53). Using

$$\frac{d\rho}{dt}=-4\rho\frac{\dot a}{a},$$

(eq. I-4) we find finally

$$\frac{d}{dt}\langle v^2\rangle=0. \qquad (58)$$

This equation says that as long as the Universe is radiation dominated and viscous dissipation can be ignored any primeval turbulence velocity is preserved, to be released, and perhaps turned to galaxy formation, when matter and radiation decouple. The Primeval Fireball is an essential element in this story for two reasons. We have shown that when pressure may be ignored, as in hypersonic turbulence, peculiar velocities vary as $v \propto a(t)^{-1}$, so the further back in time we can trace the turbulent matter-dominated phase the greater the escalation of the turbulent velocity, the more extreme the turbulence.[3] Second, hypersonic turbulence ought to dissipate itself in shocks within a few eddy turn-over times. The attractive feature in the Fireball picture is that at $z \gtrsim 1000$ the turbulence is subsonic if it is non-relativistic.

Despite these points there are some serious clouds on the horizon. Until $z \sim 10000$, when the plasma may recombine, the major dissipation of turbulence would be due to radiation diffusion through the plasma. The important number here is the ratio of eddy size λ to photon mean free path. A convenient way to specify λ is in terms of the nucleon mass $\mathfrak{M}(\lambda)$ contained within a sphere of diameter λ. This gives

$$\lambda = \frac{190\,(\mathfrak{M}/\mathfrak{M}_\odot)^{1/3}}{(\Omega h^2)^{1/3}\,(1+z)}\ \mathrm{pc}. \tag{59}$$

When this is set equal to the photon mean free path

$$\lambda_T = (\sigma_T\, n_e)^{-1} \tag{60}$$

it fixes the characteristic mass that, at epoch z, would be just optically thick for Thomson scattering,

$$\mathfrak{M}_T = \frac{1.2 \times 10^{25}\,\mathfrak{M}_\odot}{(\Omega h^2)^2\,(1+z)^6}, \tag{61}$$

assuming the matter is fully ionized. When the characteristic mass of an eddy is less than $\mathfrak{M}_T$ the radiation does not follow the eddy, and the dissipation time due to radiation drag (eq. 12) satisfies

$$\frac{t|\dot{v}|}{v} = \frac{4\,\sigma_T\, aT^4 t}{3m_p c}$$

$$\cong \frac{1.5 \times 10^{-6}\,(1+z)^{5/2}}{(\Omega h^2)^{1/2}} \tag{62}$$

$$\cong \frac{45}{(\Omega h^2)^{1/2}}, \quad z = 1000\,.$$

Because this is a large number, radiation drag at $z \sim 1000$ eliminates turbulence on a mass scale at least as large as $\mathfrak{M}_T$ (eq. 61), where $\mathfrak{M}_T = 10^7$ to $10^{11}\mathfrak{M}_\odot$, depending on the density parameter (Ωh^2).

What happens past $z \sim 1000$ depends on whether or not the plasma recombines. If it does the next question is whether the eddy turn-over time is less or greater than the expansion time t. If the latter the "turbulence" is little different from a linear perturbation, $\delta\rho/\rho < 1$. If the former the problem is that the turbulence is hypersonic, so it presumably dissipates itself in shocks in a few eddy turn-over times, the shocks starting from the scale of $\mathfrak{M}_T \sim 10^7$ to $10^{11}\mathfrak{M}_\odot$. The problem with this is that after the turbulence is dissipated the matter ends up piled up in lumps much denser than galaxies.[24]

The plasma may well be expected not to recombine because of the high turbulence energy dissipation. In this case the decoupling and associated turbulent dissipation is gentler, and conceivably the matter would end up piled in lumps of "reasonable" density. The radiation drag remains important until $t|\dot{v}/v|$ falls below unity, which according to (62) happens at redshift

$$1 + z_c \cong 200\,(\Omega h^2)^{1/5}\,. \tag{63}$$

This agrees with the epoch of decoupling of matter and radiation found in Sec. iv. When (63) is substituted into (61) it gives

$$\mathfrak{M}_T \sim \frac{1.2 \times 10^{11}\,\mathfrak{M}_\odot}{(\Omega h^2)^{16/5}}\,. \tag{64}$$

This is an estimate of the scale up to which radiation drag strongly attenuates the primeval turbulence velocity. It is just comparable to the mass of the Galaxy if $\Omega h^2 \sim 1$; and of course it is very much larger in a low density model, $\Omega h^2 \sim 0.01$.

It is hard to arrange to have turbulence with eddy size less than (60) or (61) because we do not know how it could be driven. The radiation cannot supply the needed pressure because the photon mean free path is larger than the eddy size. Matter pressure seems inadequate – for example, the turbulent velocity required to move material over the distance λ_T (eq. 60) in a characteristic expansion time $t(z)$ at redshift z_c is

$$v_c \sim \frac{\lambda_T(z_c)}{t(z_c)} \sim 2 \times 10^3 \, (\Omega h^2)^{4/5} \text{ km sec}^{-1}, \qquad (65)$$

while the thermal velocity available from matter pressure is

$$v_{th} \sim (kT_e/m_p)^{1/2} \sim 0.1 \, T_e^{1/2} \text{ km sec}^{-1}. \qquad (66)$$

The matter temperature needed to make these two velocities comparable is $T_e \sim 10^8$ K, and we do not know how the matter could be kept this hot. Thus it appears that if turbulence on a scale $>> \lambda_T$ tried to drive energy into smaller eddies it could only make shocks, and the problem again is that shocks tend to make lumps rather denser than we think we want for galaxies. For example, at redshift z_c (eq. 63) the mean nucleon density is

$$n_c \sim 100 \, (\Omega h^2)^{8/5} \text{ cm}^{-3},$$

compared with a typical density within the Galaxy,

$$n_G \sim 1 \text{ cm}^{-3}.$$

Evidently n_c is large, and, when we recall that the overdensity at a shock with "reasonable" temperature, like 10^4 K, would be substantial, we see that dissipation by shocks is liable to make objects much denser than galaxies. There could of course be such dense objects. However, this does not seem to be the way to make galaxies.

To summarize, we have argued that primeval turbulence of modest strength, such as might not be expected to dissipate in hypersonic shocks, would be dissipated by radiation drag up to the mass scale of the large galaxies (eq. 64) if $(\Omega h^2) \sim 1$. It may just be conceivable therefore that the rotation of the Galaxy, $\Theta \sim 200$ km sec^{-1}, is a primeval effect, but it is very hard to see how the rotation of the smaller galaxies could be a fossil of primeval turbulence.

It would be quite possible to make a much more complete analysis of the consequences of primeval turbulence, including effects like the details of matter recombination and collisional ionization associated with the dissipation of the turbulence, and the possible effect of a primeval magnetic field. We very much need a close and detailed analysis of this sort.

c) *Thermal History – Perturbation of the Fireball Spectrum*

Since the Universe evidently is not in statistical equilibrium at 2.7 K the Primeval Fireball radiation cannot have a strictly blackbody spectrum. One would very much like to be able to detect the deviation from a blackbody distribution and use it as indirect evidence on how the Universe has evolved. It should be emphasized once again that this is a complicated and uncertain task because there are so many undetermined or unknown effects that could be operating. This is not to say that it is a waste of time to study models for the evolution of the young Universe, only that one should be very sure to bear in mind that any model devised by man can fail in many ways. Indeed by close study of the observational consequences of the naive models and how they fail we might hope to learn whether they at least hold together reasonably well, or whether there are manifest effects we have neglected. The goal of this section will be to describe a few of the proposed lines of argument which, by virtue of their simplicity, seem to be of reasonably broad interest.

i) *Annihilation of Thermal Electron Pairs*

In the Big Bang model the early Universe was hot enough to have been flooded with thermal particle-antiparticle pairs. The last to have recombined would be the electron-positron pairs, at $T \sim m_e c^2/k \sim 10^{10}$ K, $z \sim 3 \times 10^9$. This epoch is of considerable interest because an observable amount of helium might have been produced then, as described in Chapter VIII. It suffices here to note some orders of magnitude. When $T \sim 10^{10}$ K the electron pair density is comparable to the photon number density, both being roughly $(kT/hc)^3$. The annihilation cross-section for an electron being $\sim \sigma_T$ when the energy is not too high, the mean free life of an electron at $T \sim 10^{10}$ K is given by the extrapolation of line TM in Figure VII-1 to $z \sim 3 \times 10^9$. It is apparent that the Universe ought to reach $T = 10^9$ K with the thermal electron pairs very thoroughly recombined, and the matter and radiation very close to thermal equilibrium, unless the estimate of the expansion rate is very far wrong.

ii) *Expanding from* $z = 10^8$ *to* 10^3

The non-relativistic matter tends to cool with the expansion of the Universe as $a(t)^{-2}$, the radiation as $a(t)^{-1}$, and the mixture somewhere in between. In this case the symmetry arguments of Chapter V no longer guarantee that the radiation spectrum remains thermal if initially so. For $z \gtrsim 3 \times 10^4$ Thomson scattering can relax the radiation distribution function to blackbody radiation (line TR in Figure 1). The equilibrium blackbody photon density varies as T^3, while the expansion of the Universe dilutes the photon density according to the law $n_\gamma \propto a(t)^{-3}$, which decreases with increasing expansion parameter a less rapidly than T^3 because of the pulling effect of the matter, so the expansion makes excess photons which are driven down to low frequency by Thomson scattering.

For the expansion from $z \sim 3 \times 10^4$ to 1000 we can fall back on the orders of magnitude, that the net energy lost by the radiation to the matter is small, and that the difference between the matter temperature and its equilibrium value is small. The best way to state the first number is in terms of the cooling law if the expansion were reversible. Assuming pure hydrogen, fully ionized, the thermal energy per proton is

$$u = 3\,kT + \frac{aT^4}{n},$$

the pressure is

$$P = 2n\,kT + \frac{aT^4}{3},$$

and by the usual law of thermodynamics,

$$du = T\,ds - Pd\,(n^{-1}),$$

we have that the entropy per proton is

$$s = \frac{4\,aT^3}{3n} + 2k\,\ln\left(\frac{T^{3/2}}{n}\right) + \text{constant}.$$

Suppose that at epoch a_i the temperature is T_i. Then if the expansion is reversible

$$\frac{4}{3}a\left(\frac{T_i^{\,3}}{n_i} - \frac{T^3}{n}\right) = 2k\,\ln\left[\left(\frac{T}{T_i}\right)^{3/2}\frac{n_i}{n}\right], \tag{67}$$

where

$$\frac{n_i}{n} = \left(\frac{a}{a_i}\right)^3 .$$

Equation (67) may be solved by iteration. Because $aT^3 >> nk$ a reasonable first approximation is to ignore the right hand side of (67), which gives the familiar result

$$T/T_i \;\cong\; (n/n_i)^{1/3} = a_i/a\,. \tag{68}$$

When this is substituted into the right hand side of (67) the result is

$$\begin{aligned} T &\cong \frac{T_i\,a_i}{a}\left[1 - \frac{3\,nk}{4\,aT^3}\ln\,(a/a_i)\right] \\ &\cong T_i\frac{a_i}{a}\,[1 - 7.8\times10^{-9}\,(\Omega h^2)\ln\,(a/a_i)], \end{aligned} \tag{69}$$

where the second line follows from equation (1).

The second order of magnitude is the departure of the matter temperature from its equilibrium value. This effect was discussed by Weymann.[2] Taking account of the general expansion we have that the matter temperature satisfies (eq. 15)

$$\frac{dT_e}{dt} = -2\, T_e \frac{\dot{a}}{a} + \frac{4}{3} \frac{\sigma_T\, aT^4}{m_e c} (T - T_e)\,.$$

Because the coefficient in the last term is so very large we get a good approximation to the solution by setting $dT_e/dt = 0$, to get

$$\begin{aligned} T_e &\cong T \left[1 - \frac{3}{2} \frac{\dot{a}}{a} \frac{m_e c}{\sigma_T\, aT^4} \right] \\ &= T \left[1 - \frac{500(\Omega h^2)^{1/2}}{(1+z)^{5/2}} \right]. \end{aligned} \tag{70}$$

At redshift $z = 1000$ the fractional difference between T and T_e amounts to at most 2×10^{-5}.

We have then the situation that the effective radiation temperature is pulled down by about one part in 10^8 (eq. 69) by matter colder than the radiation by one part in 10^5 (eq. 70). The maximum perturbation to the radiation spectrum expressed as an equivalent thermodynamic temperature as a function of frequency is therefore at most one part in 10^5.

iii) *Recombination of the Primeval Plasma*

The next interesting effect is the expected plasma recombination at $1 + z \sim 1000$.[25] It is easy to see that under the assumptions adopted in this chapter the recombination radiation could not be observationally interesting. The energy released is $\epsilon = 13.6$ eV per atom. As this energy is reduced by the redshift factor $(1 + z) \sim 1000$ the ratio of recombination radiation energy to Fireball energy in the present Universe is

$$\frac{\epsilon\, n_0}{aT_0^4\,(1+z)} \sim 6 \times 10^{-7}\, (\Omega h^2)\,. \tag{71}$$

This recombination radiation is redshifted to $\sim 150\ \mu$ wavelength, where it is surely lost behind the infrared radiation from interstellar dust, the most significant order of magnitude here being that we know the (as yet undetected) infrared radiation from interstellar dust must have energy density six orders of magnitude larger than (71) (cf. Sec. V-b-i).

iv) *Perturbation of the Primeval Fireball-Thomson Scattering*

If there were primeval turbulence it would be strongly dissipated near $z \sim 1000$ as the photon mean free path increases, and of course the energy lost by the matter perturbs the radiation spectrum. As bound systems like stars and galaxies form they release energy through gravitational contraction or nuclear burning. If matter and antimatter are fairly close they may annihilate. A result of each process would be to make the matter substantially hotter than the radiation, and the matter consequently may perturb the radiation spectrum. This was first discussed in detail by Weymann,[2] who pointed out that the major interactions to consider would be Thomson scattering, free-free emission, and bound-free emission.

The following discussion of Thomson scattering is based on the work of Zeldovich and Sunyaev.[13] If the plasma is much hotter than the radiation, the radiation sees electrons with mean square velocity $\langle v^2 \rangle >> kT/m_e$. In the dissipation of primeval turbulence the radiation again may see a high electron velocity dispersion $\langle v^2(t) \rangle$ as it diffuses through the matter. The result in each case is that the effective radiation temperature in the long wavelength part of the spectrum is pulled down by the factor (eq. 32)

$$T = T^* e^{-2y} ,$$

$$y = \int_{t_i}^{t} \sigma_T n_e \langle v^2 \rangle dt/3c , \qquad (72)$$

and the rate of gain of radiation energy from the matter is, by (28),

$$\frac{dU}{dt} = \frac{4}{3} \sigma_T n_e \frac{\langle v^2 \rangle}{c} U ,$$

which with (72) gives

$$U = U^* e^{4y}. \tag{73}$$

As the observed Rayleigh-Jeans temperature T_0 is reduced from what it would have been in the absence of Thomson scattering by the factor e^{2y_0} the actual radiation energy density is larger than the energy density of blackbody radiation at temperature T_0 by the factor

$$\frac{U_0}{aT_0^4} = e^{12y_0}.$$

One can find an upper bound on y_0 from the details of the spectrum measurements. It is convenient here to write the radiation frequency in terms of the variable

$$x = \frac{hc}{kT_0\lambda}, \quad \nu(t) = \frac{kT_0x}{h}\frac{a_0}{a(t)},$$

where $T_0 = 2.7$ K, and λ is the observed radiation wavelength. As in equation (IV-10) this automatically takes account of the cosmological redshift. Still assuming $T_e >> T$, we find that equation (25) with (72) is

$$\frac{\partial \mathfrak{N}}{\partial y} = x^2 \frac{\partial^2 \mathfrak{N}}{\partial x^2} + 4x \frac{\partial \mathfrak{N}}{\partial x}, \tag{74}$$

with the initial value

$$\mathfrak{N}_i = [\exp(xe^{-2y_0} - 1)],$$

where y_0 is the result of extending the integral in (72) over the whole interval of time when $T_e >> T$. The solution to the linear equation (74) has been discussed by Sunyaev and Zeldovich,[13] who conclude

$$y_0 \lesssim 0.1, \tag{75}$$

for otherwise the radiation spectrum fitted to the microwave flux measurements would conflict with the CN measurement.

According to (73) and (75) the original fireball energy might have been increased by as much as a factor $e^{0.4} = 1.5$. If this added energy came from primeval turbulence deposited in the Fireball radiation at redshift z the energy requirement is

$$a(T_0 e^{2y_0} (1+z))^4 (e^{4y_0} - 1) = \frac{1}{2} \langle v^2 \rangle \rho_0 (1+z)^3 ,$$

which gives

$$\frac{\langle v^2 \rangle^{1/2}}{c} \lesssim 0.007 (1+z)^{1/2} (\Omega h^2)^{-1/2} .$$

If $z = 1000$ this means $v/c \lesssim 0.2 (\Omega h^2)^{-1/2}$, which velocity is impressively high but perhaps possible in a dense strong turbulence model.

Another way to express the limit on y_0 is in terms of an assumed electron temperature, maintained perhaps by matter-antimatter annihilation. If the electron temperature were constant at T_e since z_m then

$$y_0 = \int_{t_m}^{t_0} \frac{\sigma_T n_e kT_e dt}{m_e c}$$

$$\cong \frac{2}{3} \frac{\sigma_T n_e(t_0) kT_e (1+z_m)^{3/2}}{m_e c H \Omega^{1/2}} .$$

Assuming the matter is fully ionized the limit $y_0 \lesssim 0.1$ gives

$$T_e \lesssim \frac{1.3 \times 10^{10} \text{ K}}{(1+z_m)^{3/2} (\Omega h^2)^{1/2}} , \tag{76}$$

and if z_m were the nominal epoch of decoupling, $z_m \sim 1000$, the limit would be

$$T_e \lesssim 4 \times 10^5 (\Omega h^2)^{-1/2} \text{ K} .$$

If the plasma had been hotter than this at $z \sim 10^3$ it would have caused a perturbation to the Fireball spectrum larger than the observational limits would permit.

v) *Bremsstrahlung Radiation*

A second interesting effect is the contribution to the microwave background by bremsstrahlung radiation from a plasma in the early Universe, at high redshift. This process has already been discussed in connection with the X-ray and ultraviolet radiation background in Section IV-d-iv. The differences here are (1) radio emission is only interesting at high density, hence high redshift; (2) the frequency is much lower so we have to worry about self-absorption and stimulated emission. The radio background as a test of a plasma at high redshift was first discussed by Kaufman,[26] who considered the assumption that the plasma temperature was constant in time. Kaufman concluded that if the plasma temperature were $T \sim 10^5$ K and if Ωh^2 were ~ 1 the resulting flux at 7 cm wavelength might agree with the microwave background that had just been discovered by Penzias and Wilson, and that if T were any larger Ωh^2 would have to be less than unity. Since then the background spectrum at somewhat longer wavelength has been mapped out, and this gives a better limit. The test was reconsidered by Weymann,[2] who pointed out that an interesting wavelength for a test for free-free emission at high redshift would be $\sim$ 20-30 cm. Sunyaev[27] discussed the effect in the following convenient model. Suppose that at epoch z_m the uniformly distributed matter was ionized and maintained thereafter at temperature T_e much greater than the radiation temperature. Then the ratio of radio brightness due to free-free emission to brightness of the Primeval Fireball is (eq. IV-12)

$$\frac{i_{ff}}{i} = \frac{T_e}{T_0} \int_{t_m}^{t_0} \kappa(\nu(t))\, c\,dt\, \frac{a(t)}{a_0}\,, \tag{77}$$

where T_0 is the present radiation temperature and $\nu(t) = \nu_0\, a_0/a(t)$, ν_0 being the observed frequency. Over the range of the observed microwave background spectrum there is no very large deviation from a thermal spectrum, so under the condition $T_e >> T$ it is obvious that self-absorption could not be important. The condition on (77) is

$$i_{ff} \lesssim i\,. \tag{78}$$

For frequencies of interest here $h\nu/kT << 1$, and equation IV-65 for free-free opacity has to be multiplied by a correction factor $g \lesssim 6$ if $T \lesssim 10^6$, $z_m \gtrsim 100$ (ref. IV-51). We have then from (77) and (78)

$$(1 + z_m) \lesssim \frac{2.0 \times 10^4 \, T_4^{1/3}}{\lambda^{4/3} \, (\Omega h^2)} , \tag{79}$$

where T_4 is in units of 10^4 K. The longest observed wavelength might be taken to be $\lambda = 50$ cm, the shorter of the two wavelengths used by Howell and Shakeshaft (Table V-1), which gives

$$1 + z_m \lesssim \frac{110 \, T_4^{1/3}}{(\Omega h^2)} . \tag{80}$$

It follows that *if* $\Omega h^2 \sim 1$ *and if* the plasma had been reasonably cool, $T_4 \sim 1$, then the Universe must have enjoyed an interval between redshifts $z = 1000$ and $z \sim 300$ when the uniformly distributed matter was neutral, or else at the same temperature as the radiation.

Zeldovich and Sunyaev[13] pointed out that (76) and (80) together can be used to place an upper bound on z_m for any plasma temperature. The temperature for maximum z_m consistent with both equations is gotten by setting (76) equal to (80), which gives

$$T_4 \sim 110 \, (\Omega h^2)^{2/3} ,$$

and the upper bound on redshift is then

$$(1 + z_m) \lesssim 530/(\Omega h^2)^{7/9} .$$

This equation suggests that a dense gas must have enjoyed a cool period, although it will be noted that the limit is not all that strong. For example even if $\Omega = 1$ the limit would exceed $z = 1000$ if h were 0.5 ($H = 50$ km sec^{-1} Mpc^{-1}).

REFERENCES

1. A review of some aspects of this question is given by R. A. Sunyaev and Ya. B. Zeldovich, *Comments in Astrophysics and Space Physics* 2, 66, 1970.
2. Histories of an intergalactic medium have been discussed by R. Weymann, *Ap. J.* 145, 560, 1966 and *Ap. J.* 147, 887, 1967; V. L. Ginzburg and L. M. Ozernoi, *Astron. Zh.* 42, 943, 1965; Engl. tr. *Soviet Astronomy –A.J.* 9, 726, 1966; M. J. Rees and D. W. Sciama, *Ap. J.* 145, 6, 1966; R. J. Gould, *Ann. Rev. Astron. Astrophys.* 6, 195, 1968; J. Arons and R. McCray, *Astrophysical Letters* 5, 123, 1969; J. Bergeron, *Astron. and Astrophys.* 3, 42, 1969; M. J. Rees and G. Setti, *Astron. and Astrophys.* 8, 410, 1970.
3. F. Hoyle, Solvay Conference on the *Structure and Evolution of the Universe*, 1958.
4. E. N. Parker, *Ap. J.* 160, 383, 1970.
5. K. S. Thorne, *Ap. J.* 148, 51, 1967.
6. e.g. H. Alfvén, ref. I-37; R. Omnès, *Phys. Rev. Letters* 23, 38, 1969; *Astron. and Astrophys.* 10, 228, 1971.
7. E. R. Harrison, *Phys. Rev. Letters* 18, 1011, 1967.
8. S. L. Jaki, The Paradox of Olbers Paradox (Herder and Herder, N. Y., 1969).
9. W. J. Boardman, *Ap. J. Suppl.* 9, 185, 1964.
10. R. Weymann, *Phys. Fluids* 8, 2112, 1965.
11. L. Spitzer, *Physics of Fully Ionized Gases* (2nd ed., Interscience, N. Y.) p. 135.
12. A. S. Kompaneets, *Zh. E.T.F.* 31, 876, 1956; Engl. tr. *Soviet Phys.-J.E.T.P.* 4, 730, 1957.
13. Ya. B. Zeldovich and R. A. Sunyaev, *Astrophysics and Space Science* 4, 301, 1969.
14. The weak-field equations are given by Tolman, ref. VI-2, §93; the gravitational field equations are written down in §78, and the matter stress-energy tensor of the sort we are considering in §85.

15. E. M. Lifshitz, *J. Phys.* 10, 116, 1946; W. B. Bonnor, *M.N.* 117, 104, 1957.
16. P. J. E. Peebles and R. H. Dicke, *Ap. J.* 154, 891, 1968; P. J. E. Peebles, *Ap. J.* 155, 393, 1969.
17. A. G. W. Cameron, *Icarus* 1, 13, 1962; Ya. B. Zeldovich, *Astron. Zh.* 46, 775, 1969; Engl. tr. *Soviet Astronomy-A.J.* 13, 608, 1970.
18. P. J. E. Peebles, *Ap. J.* 142, 1317, 1965.
19. P. J. E. Peebles, *Ap. J.* 147, 859, 1967.
20. C. W. Misner, *Ap. J.* 151, 431, 1968.
21. C. F. von Weizsäcker, *Ap. J.* 114, 165, 1951.
22. G. Gamow, *Phys. Rev.* 86, 251, 1952; *Proc. Nat. Acad. Sci.* 40, 480, 1954.
23. J. Oort, *Nature* 224, 1158, 1969; L. M. Ozernoi and A. D. Chernin, *Astron. Zh.* 45, 1137, 1968; Engl. tr. *Soviet Astronomy-A.J.* 12, 901, 1969; E. R. Harrison, *M.N.* 147, 279, 1970; H. Sato, T. Matsuda and H. Takeda, *Progr. Theor. Phys.* 43, 1115, 1970.
24. P. J. E. Peebles, *Astrophysics and Space Science.* 10, 280, 1971.
25. P. J. E. Peebles, *Ap. J.* 153, 1, 1968; Ya. B. Zeldovich, V. G. Kurt and R. A. Sunyaev, *Zh. E.T.F.* 55, 278, 1968, Engl. tr. *Soviet Phys.-J.E.T.P.* 28, 146, 1969.
26. M. Kaufman, *Nature* 207, 736, 1965.
27. R. A. Sunyaev, *Doklady USSR* 179, 45, 1968; Engl. tr. *Soviet Phys.-Doklady* 13, 183, 1968.

VIII. PRIMEVAL HELIUM

This is an important subject because it is by far the most powerful probe we have on what might have been happening in the distant past when, according to the Big Bang Primeval Fireball picture, the Universe was dense and hot. The rather lengthy computation of primeval helium production may be summarized in three steps: (1) When $t < 1$ sec, time being measured from the singularity $a = 0$, $\rho = \infty$ in the Big Bang, the temperature is greater than 10^{10} K, and matter and radiation are expected to have been very close to thermal equilibrium. The significant equilibrium constituents at 10^{10} K are free protons, neutrons, electrons, positrons, neutrinos and radiation. (2) At $T \sim 10^{10}$ K, $t \sim 1$ sec, equilibrium is broken with the freezing-in of the neutron abundance at $n/p \sim 0.1$. (3) At $T \sim 10^{9}$ K, $t \sim 300$ sec, the remaining neutrons react with protons to form deuterium, which burns through to helium.

In the original Gamow-Alpher theory, as outlined in Chapter V, it was supposed that the matter initially was all or in large part neutrons, given as an initial condition (ref. V-3). Working from this initial condition Fermi and Turkevich[1] verified by detailed numerical computation of the nuclear reaction rates Alpher's original observation, that it is easy to get helium but it is hard to see how the process could carry past the mass 5 gap to make any appreciable amount of heavier elements. Hayashi[2] pointed out that the neutron abundance is not a free initial condition, that this parameter is determined by reactions like inverse beta decay in the still earlier Universe. Taking up this idea, in 1953 Alpher, Follin and Herman computed in detail the expected time variation of the neutron abundance.[3]

On the observational side there were some hints that the picture of helium production in the Big Bang may be valid. In 1961 Osterbrock and

Rogerson pointed out that the similarity of element abundances in the Orion Nebula, representing present day interstellar material, and in the Sun, material isolated some 4.6×10^9 y ago, may indicate that helium in particular was produced before the Galaxy in the Big Bang.[4] In 1964 O'Dell, Peimbert and Kinman presented abundance determinations for a planetary nebula in the globular star cluster M15, a representative of the oldest known stellar population.[5] The oxygen abundance was markedly deficient relative to the Sun, as is perhaps expected if the material is closer to the primeval composition, but the helium abundance was fully as high as the present "cosmic" value. Of course the material in a planetary nebula comes from a highly evolved and excited star, so there was the clear possibility that this helium was produced in the star, but it was pointed out that it could also mean that the primeval helium abundance was high. In the same year Hoyle and Tayler[6] discussed in some detail the remarkable uniformity in the helium abundance. They pointed out that this could be understood in several ways. One is the Gamow-Alpher picture of helium production in the Big Bang. Another may be helium production in supermassive stars of some sort in an early phase of the formation of the Galaxy.

The next year with the discovery of the candidate for the Primeval Fireball (ref. V-7) Peebles independently hit on helium production as an application of Dicke's Fireball picture, and computed in some detail the primeval helium production with parameters fixed by this candidate.[7] Wagoner, Fowler and Hoyle repeated the calculation and also obtained, by detailed attention to all reasonably possible reactions, the very small abundances of elements heavier than helium that would be produced in the Big Bang.[8] The current interest in the primeval helium problem is based on these two points – first that there is accumulating (but not unequivocal) evidence for uniform helium abundance, suggesting that the helium was produced not in the diverse observed objects but in some universal process before the objects formed; and second that the Primeval Fireball seems to be evidence that the Universe really did expand from a hot dense state, and it fixes values for parameters conducive to appreciable but not necessarily excessive helium production.

There are two points of sharp debate on this subject. (1) Is the initial primeval helium abundance of the galaxies high, 25-30% by mass, as suggested in the "naive Big Bang"? (2) Is the naive Big Bang so excessively naive, with so little account of all the complexity of the real world, as to be irrelevant? It will be argued in the following sections that the answers are (1) we don't know for sure; and (2) not necessarily.

a) *Helium Production in the Naive Big Bang*

i) *Some Assumptions*

The calculation is based on the assumption that the Universe did expand away from a hot dense phase of the Universe, from temperature $T \gtrsim 10^{12}$ K, and that the Universe was homogeneous and isotropic in all detail during the helium production epoch. This is not directly indicated by the Primeval Fireball, because we do not need to go back to such high temperatures to be assured of a thermal radiation spectrum. It is strictly a semi-empirical extrapolation, and we can readily imagine modifying it in relatively small detail, like keeping the radiation density uniform but making the matter distribution irregular, or in large, by assuming a dense but irregular chaotic Universe, or simply by denying that the Universe ever was this dense and hot. The first justification for the extrapolation is that we owe it to the classical cosmologists to follow in detail the straight consequences of their models, to see whether there is objective evidence that these models may have some validity, or must be made more general, or must be abandoned. The second justification is an instability argument that says irregularities tend to grow worse as the Universe evolves. One example was given in Section VII-b, for irregularities on the scale $\lambda << ct$. Pressure can stabilize irregularities up to the Jeans length $\lesssim ct$, but beyond this there is no known stabilization for density irregularities, and one can in fact find explicit solutions in which, aside from special choices of initial values, the Universe grows ever more irregular as time goes on (ref. VII-19). This is not a universally accepted view (ref. VII-18), but unless and until we have an explicit counter-example showing what the stabilization mechanism is, it merits serious attention. It would say that, as the Universe is fairly homogeneous and regular now, it must have been highly regular in the past. One should of course bear in mind that this is not a proof, so we must be prepared to consider other possibilities.

Further assumptions, rather in the nature of parameter choices, will be adopted in the "naive Big Bang" calculation:

1) General Relativity is valid: it is daring to extrapolate so far such an inadequately tested theory. On the other hand the formula we use for the expansion rate is just

$$\frac{\dot{a}^2}{a^2} = \frac{8}{3}\pi G\rho \ , \tag{1}$$

and, as Gamow emphasized, this just says that the kinetic energy in expansion almost exactly balances the gravitational potential energy, which seems so simple as to be convincing. The major difficulties one might imagine are that G might vary, or that there may be new contributions to the total mass density, as in the scalar-tensor theory of Brans and Dicke.[9]

2) No new neutrinos: we assume that the conventional understanding of the laws of physics at energies ~ 1 MeV and nucleon densities ~ 1 g cm^{-3} is valid and may be applied even in the very distant past.

3) There is a Primeval Fireball, present temperature $T_0 = 2.7$ K.

4) No neutrino degeneracy: if the Universe contains enough electron-type neutrinos, or else antineutrinos, they may be degenerate, Fermi energy $> kT$, and this will alter the equilibrium neutron abundance, through reactions like inverse beta decay. It is assumed in this Section that this is not the case. This implies a neutrino or antineutrino Fermi energy $<< (2.7\ \mathrm{K})k$ now; equivalently $<< 10^8$ neutrinos (or antineutrinos) per nucleon, which seems liberal.

5) No matter – antimatter annihilation: the nucleon density at epoch t is assumed to be fixed by the equation

$$n = n_0\, a_0^3/a(t)^3 \ , \tag{2}$$

where n_0 is the present value (eq. IV-1),

$$n_0 \equiv 1.12 \times 10^{-5}\, \Omega h^2\ \text{cm}^{-3},\ 0.01 \lesssim \Omega h^2 \lesssim 1. \qquad (3)$$

This assumes matter and antimatter have not been annihilating to any very great extent.

6) No primeval magnetic field: if there were a general intergalactic magnetic field of about 10^{-6} Gauss, like the field within the Galaxy, then the extrapolation back in time, based on flux conservation, gives at 10^{10} K

$$B \sim 10^{-6}\,(10^{10}/2.7)^2 \sim 10^{13}\ \text{Gauss},$$

strong enough to perturb the single-particle electron energy levels, which in turn affects the rate of relaxation of the neutron-proton abundance ratio.[10] Matese and O'Connell argue that the net effect of a strong primeval field would be to increase the primeval helium production.

Each one of these assumptions is subject to debate, and some consequences of varying them will be discussed in Section b below. In addition to these assumptions there are as usual a number of computational approximations, as will be revealed.

ii) *Conditions at* $T = 10^{11}$ *to* 10^{12} K; *Thermal Equilibrium*

Even with all the above assumptions it is not obvious ahead of time that we can compute the primeval abundances of the elements, for how can we understand the state of matter issuing from the singularity $a = 0$, $\rho = \infty$ at $t = 0$? The enormous point of simplification is that thermal relaxation is expected to have washed over all remnants of the entirely obscure start of the expansion.

At the epoch defined by Fireball temperature $T = 10^{12}$ K we have the following orders of magnitude. The typical thermal energy is

$$kT \sim 100\ \text{MeV}.$$

The nucleon density is, by (2) and (3),

$$n \sim 10^{28}\ \mathrm{cm}^{-3},$$

$$\rho_m \sim 10^4\ \mathrm{g\ cm}^{-3}.$$

The equivalent mass density in electromagnetic radiation is much larger,

$$\rho_r = \frac{aT^4}{c^2} \sim 10^{13}\ \mathrm{g\ cm}^{-3},$$

so by (1) the expansion rate is a function of temperature alone, and the characteristic expansion time scale is

$$t \sim (G\,\rho_r)^{-1/2} \sim 10^{-4}\ \mathrm{sec}.$$

It will be noted that with the exception of ρ_r these quantities, taken singly, are by no means extreme. They are well within the range of observed known phenomena, and unless things go very wrong we ought to be able to understand the properties of matter under these conditions.

The expansion time scale is long enough to permit decay of the exotic particles created in the earlier Universe, the thermal energy is high enough to evaporate (photo-dissociate) any complex nuclei, and again the time scale is long enough to do this. Thus we expect to find ordinary free neutrons and protons, and since the inter-baryon distance is large, on the order of 10^{-9} cm, the neutrons and protons should act like an ideal gas.

At $T > 10^{10}$ K the thermal energy is high enough to permit electron pair formation in reactions like

$$\gamma + \gamma \rightarrow e^+ + e^- . \tag{4}$$

Let us consider first the expected equilibrium abundance of electron-positron pairs, and then verify that the relaxation rate is fast enough to reach this equilibrium. In all the following calculations electromagnetic interactions among particles will be ignored, which means here that the electrons are assumed to act like free particles. Then the probability of finding an electron in the single particle state (mode) with energy $E(p)$ (p is the electron momentum in the state) is proportional to

$$e^{(\mu - E(p))/kT}, \tag{5}$$

where μ is the chemical potential. To determine μ we have the fact that the sums of chemical potentials entering and leaving a reaction like (4) are conserved, whence for electrons and positrons[11]

$$\mu^- + \mu^+ = 0, \tag{6}$$

because the chemical potential for blackbody radiation is zero. As will be verified there are many more electron-positron pairs than extra electrons belonging to the protons, and the near equality of electron and positron abundance gives by symmetry $\mu^- = \mu^+ = 0$. Because there can only be one electron or none in a single particle state we have that the probability that a chosen state (mode) is occupied is (cf. eq. V-7)

$$\begin{aligned} \mathfrak{N} &= e^{-E(p)/kT}/(1 + e^{-E(p)/kT}) \\ &= 1/(e^{E(p)/kT} + 1), \end{aligned} \tag{7}$$

where

$$E^2 = p^2c^2 + m^2 c^4. \tag{8}$$

The number density of single particle states is

$$g\, d^3p/(2\pi\hbar)^3, \tag{9}$$

where the factor g represents the number of spin states for given momentum – for an electron $g = 2$. By (7) and (9) the equilibrium number density of electrons plus positrons is

$$n_{th} = \frac{2}{\pi^2 \hbar^3} \int_0^\infty \frac{p^2\, dp}{e^{E/kT} + 1}. \tag{10}$$

When $kT >> m_ec^2$ the electrons are relativistic, the energy (8) may be approximated as $E = pc$, and equation (10) becomes[11]

$$n_{th} = \frac{2}{\pi^2 \hbar^3} \int_0^\infty \frac{p^2\, dp}{e^{pc/kT} + 1}$$

$$= \frac{2}{\pi^2} \left(\frac{kT}{\hbar c}\right)^3 \int_0^\infty \frac{x^2\, dx}{e^x - 1}, \tag{11}$$

$$kT >> m_e c^2.$$

It will be noted that the number density of photons in blackbody radiation differs from (11) only by the minus sign in the denominator in the constant integral (eq. V-7). That is, the thermal electron pairs act very much like radiation. The energy density and pressure contributed by the electron-positron pairs are readily found from (7) and (9) to be

$$\mathcal{E} = \frac{2}{\pi^2 \hbar^3} \int_0^\infty \frac{E(p)\, p^2\, dp}{e^{E/kT} + 1},$$

$$P = \frac{2c^2}{3\pi^2 \hbar^3} \int_0^\infty \frac{p^4\, dp\, E^{-1}}{e^{E/kT} + 1}. \tag{12}$$

In the high temperature limit these equations reduce to[11]

$$\mathcal{E} = \frac{7\pi^2}{60} \frac{(kT)^4}{(\hbar c)^3}$$

$$= \frac{7}{4} aT^4, \tag{13}$$

$$P = \mathcal{E}/3,$$

where a is Stefan's constant,

$$a = \pi^2 k^4 / 15 \hbar^3 c^3.$$

According to equation (11) the inter-particle distance for electrons is on the order of $\hbar c/kT \sim 3 \times 10^{-12}$ cm at $T = 10^{11}$ K, and the electrostatic energy of interaction of the electrons is on the order of $e^2\, kT/\hbar c$, which is smaller than the kinetic energy kT by the fine structure constant 1/137. Thus the free particle picture is a fairly reasonable approximation.

The ratio of the number density of electron pairs to excess electrons is on the order of

$$\frac{n_{th}}{n} \sim \left(\frac{kT}{\hbar c}\right)^3 / n \sim 10^8 , \tag{14}$$

nearly independent of time. The approximation in equation (7), where the chemical potential was dropped, is based on the large size of this number.

Now let us consider the rate of relaxation to this equilibrium. If kT is not too much larger than mc^2 the cross-section for electron-positron annihilation is about the same as the Thomson scattering cross-section, $\sigma_T \sim 6 \times 10^{-25}\ cm^2$, so the mean free time for annihilation of an electron and positron is

$$t_c \sim (\sigma_T\, n_{th}\, c)^{-1} \sim 10^{-21}\ sec \tag{15}$$

at $T = 10^{11}$ K. This characteristic time evidently applies as well to pair production and to thermalization of the radiation because the number densities of photons and electrons are comparable. This relaxation time is some 17 orders of magnitude shorter than the expansion time. Thus aside from malevolent choices of initial values (like all the energy concentrated in a few quanta of enormous energy) the radiation and the electrons have to relax to thermal equilibrium.

For neutrinos we cannot argue ahead of time that there should be almost identical numbers of particles and antiparticles because we have nothing like charge equality to nail down the number. The conserved quantity for electron-type neutrinos is the lepton number ℓ = number of neutrinos minus number of antineutrinos plus number of electrons minus number of positrons. Because charge neutrality keeps the last two numbers almost equal it is a question of a superabundance of neutrinos or of antineutrinos. In this Section we are assuming that $|\ell| << n$, which makes the chemical potential of the neutrinos small, much less than kT, so equations (7) and (9) apply.

As there is one spin state for neutrinos, one for antineutrinos, we have that the energy density of electron-type neutrinos is

$$\mathcal{E}_\nu = \frac{7}{8} a T_\nu{}^4 , \tag{16}$$

where T_ν is the neutrino temperature. There would be an equal contribution from muon type neutrinos. Again, the neutrino number density is comparable to the photon density.

There is some speculation that neutrinos might be produced by reactions like

$$\begin{aligned} e^+ + e^- &\rightleftarrows \nu + \bar{\nu} , \\ e^+ + \nu &\rightleftarrows e^+ + \nu , \end{aligned} \tag{17}$$

the cross-section at 10^{11} K being $\sim 10^{-42}$ cm^2. The equilibrium density of electrons at this temperature is $n_{th} = 0.18\,(kT/\hbar c)^3 \sim 10^{34}$ cm^{-3}, so the mean free time for a neutrino under the reactions (17) is ~ 0.003 sec, comparable to the expansion time, which means the reaction is just about capable of thermalizing the neutrino distribution at this temperature. Because the electron density and the cross-section are rapidly increasing functions of temperature the neutrinos clearly would have relaxed to equilibrium if we had started at a somewhat higher temperature. A second set of reactions important at temperatures slightly higher than 10^{11} K are

$$\begin{aligned} e^+ + \mu^- &\rightleftarrows \bar{\nu} + \nu_\mu , \\ \nu_\mu + \mu^+ &\rightleftarrows e^+ + \nu , \end{aligned}$$

and so on. At $T \gtrsim 10^{12}$ K muon pairs are as abundant as electrons and, as in the discussion of (17), one finds that there is time for relaxation to equilibrium.

When $T < 10^{11}$ K there is no known interaction strong enough to thermalize neutrinos, the neutrino mean free path becomes much greater than ct, and neutrinos are in effect decoupled from the matter and radiation. Just as in the discussion of blackbody radiation one readily sees that the decoupled neutrinos cool with the expansion of the Universe as $T_\nu \propto a(t)^{-1}$. When $T > 10^{10}$ K, $kT \gtrsim 1$ MeV, the electron pairs behave like radiation, so the strongly coupled matter and radiation cool according to this same law and equilibrium is not in fact broken until at 10^{10} K the electron pairs start to recombine and dump their energy into the radiation.

The simplest way to get the relation between $a(t)$ and the radiation temperature as the electrons recombine is to use the fact that the relaxation time (15) is very short, so the recombination process is reversible. The entropy of the radiation plus electron pairs in a comoving volume $V = V_0 a(t)^3$ is found from (7) and (9) to be

$$S = \frac{4}{3} a\, V_0\, (a(t)\, T)^3 \left[1 + \frac{15}{2\pi^4} \int \frac{x^2\, dx\, (x^2 + 3y^2)}{y\, (e^y + 1)}\right], \tag{18}$$

where

$$x = pc/kT\,,$$

$$y = E/kT\,,$$

E and p being the electron energy and momentum, and T the temperature of the radiation and electrons. Because S is conserved, a constant, (18) gives directly the expansion parameter a as a function of T. When $kT >> m_e c^2$, $x \cong y$, and the quantity in square parentheses in (18) is just $[1 + 7/4]$. When kT falls well below $m_e c^2$ the integral becomes very small, the quantity in square parentheses is just unity, so $(aT)^3$ has increased by the factor $(11/4)$. Because the neutrino temperature is proportional to $a(t)^{-1}$ and started out the same as the radiation temperature the neutrinos end up colder than the radiation by the factor

$$T_\nu / T_e = (4/11)^{1/3}.$$

The neutron-proton abundance ratio is fixed by reactions like $e^- + p \rightleftarrows n + \nu$. As we are assuming that the chemical potentials of the electrons and neutrinos are negligibly small the chemical potentials of neutrons and protons are equal, and the equilibrium abundance ratio is by (5) simply

$$\frac{n}{p} = e^{-Q/kT} ,$$

$$Q = (m_n - m_p)c^2 . \tag{19}$$

The rate of relaxation to this equilibrium is taken up in the next Section, where it is shown that n/p is frozen in at $n/p \sim 0.1$. It is these residual neutrons that make helium production possible.

iii) *Neutron-Proton Abundance Ratio*

The neutron-proton abundance ratio is fixed by the reactions

$$\begin{aligned} n &\rightleftarrows p + e^- + \bar{\nu} , \\ e^+ + n &\rightleftarrows p + \bar{\nu} , \\ \nu + n &\rightleftarrows p + e^- . \end{aligned} \tag{20}$$

The rates of these reactions may be computed in first order perturbation theory, and the strength of the interaction determined from the known free neutron decay rate,

$$n \rightarrow p + \bar{\nu} + e^- . \tag{21}$$

In the standard way we imagine the decaying neutron is in a box with volume V, and fix the (free) wave functions to be periodic in V. The rate for the reaction (21) is given by the usual time dependent first order perturbation theory result,

$$\lambda = \frac{\ln 2}{t_{1/2}} = \sum \frac{2\pi}{\hbar} |\langle H \rangle|^2 \, \delta(Q - E_e - E_\nu) , \tag{22}$$

where the net decay energy is

$$Q = (m_n - m_p)c^2 = 1.293 \text{ MeV} . \tag{23}$$

The term $|\langle H \rangle|^2$ represents the (constant) interaction energy matrix element (averaged over particle spins), and the sum is over all the possible e and ν states in V. This sum may be replaced with an integral, using the density of states (9),

$$\frac{\ln 2}{t_{1/2}} = \frac{2\pi}{\hbar} |\langle H\rangle|^2 \int \frac{V\, d^3 p_\nu}{(2\pi\hbar)^3} \cdot \frac{2V\, d^3 p_e}{(2\pi\hbar)^3}\, \delta(Q - E_e - E_\nu)$$

$$= \left\{\frac{2\pi}{\hbar} |\langle H\rangle|^2\right\} \cdot \frac{2(4\pi)^2 V^2}{(2\pi\hbar)^6 c^3} \int (Q - E_e)^2 p_e^2\, dp_e \,. \tag{24}$$

There is a factor of 2 for the two electron spin states, while the neutrino has only one spin. It is conventional to express the integral in (24) as

$$\int (Q - E_e)^2 p_e^2\, dp_e \equiv f\, m_e^5 c^7 \,,$$

where the dimensionless factor f is

$$f = 1.634 \,. \tag{25}$$

This is based on free wave functions, neglecting the Coulomb perturbation. When this is taken into account f becomes 1.689, 3 percent larger than (25), but as Coulomb corrections are being ignored it seems reasonable to be consistent about it. Collecting, we have the desired matrix element in terms of the measured half-life,

$$\frac{2\pi}{\hbar} |\langle H\rangle|^2 = \frac{2\pi^4 \hbar^6 \ln 2}{V^2 f\, t_{1/2}\, m_e^5 c^4}. \tag{26}$$

The most recent measurement of the half-life gave[12]

$$t_{1/2} = 10.80 \pm 0.16 \text{ minutes}, \tag{27}$$

where the error is a standard deviation. Although parts of the interaction are known from nuclear beta decay (27) still seems to be the best way to fix (26).[13]

To find the cross-section for the reaction $\nu + n \to p + e^-$ one again imagines that the two particles ν and n are in V. Then (22) gives the reaction rate as

$$\frac{\sigma c}{V} = \left\{\frac{2\pi}{\hbar} |\langle H\rangle|^2\right\} \int \frac{2V\, d^3 p_e}{(2\pi\hbar)^3}\, \delta(Q + E_\nu - E_e) \,, \tag{28}$$

which with (26) gives

$$\sigma = \frac{2\pi^2 \hbar^3 \ln 2 \, v_e (E_\nu + Q)^2}{ft_{1/2} \, m_e^5 \, c^9}, \tag{29}$$

where in evaluating the integral one uses

$$p_e^2 \, dp_e = v_e E_e^2 \, dE_e/c^4 ,$$

$v_e = p_e c^2/E_e$ being the electron velocity.

Before computing in any more detail it is useful to examine the orders of magnitude. When $T = 10^{10}$ K the neutrino energy is $E_\nu \sim kT = 0.86$ Me and with (25) and (27) the cross-section (29) amounts to

$$\sigma \cong 4 \times 10^{-43} \text{ cm}^2 . \tag{30}$$

Also, at $T = 10^{10}$ K the number density of neutrinos is (eqs. 7, 9, 11)

$$\begin{aligned} n_\nu &= \frac{1}{2\pi^2}\left(\frac{kT}{\hbar c}\right)^3 \int_0^\infty \frac{x^2 \, dx}{e^x + 1} \\ &= 0.091 \, (kT/\hbar c)^3 = 7 \times 10^{30} \text{ cm}^{-3} . \end{aligned} \tag{31}$$

The rate of the reaction $\nu + n \to p + e^-$ is roughly equal to the product of the cross-section (30) with the neutrino density (31) and with the velocity of light,

$$\sigma \, n_\nu \, c \cong 0.1 \text{ sec}^{-1} . \tag{32}$$

There is a similar contribution to the rate of conversion of neutrons to protons by positron capture, and of course when the neutron-proton abundance ratio is equal to the equilibrium value (19) this is balanced by the net rate going the other way. It will be noted that (32) increases very rapidly with increasing temperature, a factor T^3 coming from the neutrino density and a factor T^2 from the cross-section because σ varies roughly as the square of the energy.

The rate to compare (32) with is the rate of expansion of the Universe, which since $\rho \sim aT^4/c^2$ is (eq. 1)

$$\frac{\dot{a}}{a} = \left(\frac{8}{3}\pi G_\rho\right)^{1/2} \sim 0.2 \sec^{-1} . \tag{33}$$

Equations (32) and (33) say that at $T = 10^{10}$ K the rate at which a neutron is transformed to a proton is just comparable to the rate at which the Universe is expanding and cooling. Because (32) is a rapidly increasing function of temperature, we conclude that the reactions (20) are fast enough to lock the n/p abundance ratio to the equilibrium value (19) when $T > 10^{10}$ K, but the neutron abundance freezes in at a temperature T^* near 10^{10} K. The residual frozen-in neutron abundance is then

$$n/p = e^{-Q/kT^*} \sim e^{-1.5} = 0.2 .$$

It is a fascinating coincidence of the orders of magnitude for expansion rate and inverse beta decay rate that the neutron abundance is frozen in when it has just become a sensitive function of temperature, n/p neither negligibly small nor closely equal to 0.5, so we have an interesting neutron abundance from which to make helium, and at the same time the helium abundance can be a sensitive test of the details of the theory of this early phase of the Universe.

To get a more accurate reaction rate for $n + \nu \to p + e^-$ we have to take account of the fact that some of the electron states are already occupied, and we have to integrate over the neutrino energy distribution. By (7) the fraction of the single electron states at energy E_e not already occupied is

$$1 - \mathfrak{N} = \left(1 + e^{-E_e/kT}\right)^{-1} . \tag{34}$$

The number density of neutrinos with momentum in the range p_ν to $p_\nu + dp_\nu$ is

$$d\, n_\nu = \frac{p_\nu{}^2\, dp_\nu}{2\pi^2 \hbar^3}\left(e^{p_\nu c/kT_\nu} + 1\right)^{-1} . \tag{35}$$

The net reaction rate per neutron is the product of the cross-section (29), the fraction (34) of unocccupied states, the neutrino density (35), and the neutrino velocity c, integrated over p_ν,

$$\langle \sigma\, nc \rangle = \frac{\ln 2}{ft_{1/2}\, m_e{}^5\, c^8} \int \frac{p_\nu{}^2\, dp_\nu\, v_e\, E_e{}^2}{\left(e^{p_\nu c/kT_\nu} + 1\right)\left(1 + e^{-E_e/kT}\right)}, \qquad (36)$$

where

$$Q + p_\nu c = E_e .$$

The rates of each of the other reactions (20) may be found in the same way, and are listed in references (3), (5) and (6).

Now we are in a position to find the time variation of the neutron abundance when heavier element production may be neglected. The expansion parameter is given as a function of time by (1), where the important contributions to ρ are the radiation energy density aT^4, the energy density (16) $\frac{7}{4} aT_\nu{}^4$ for neutrinos of the muon type and the electron type, and the energy density (12) of electron pairs. The radiation temperature is given

TABLE VIII-1

NEUTRON ABUNDANCE

T^+	$T_\nu{}^+$	t(sec)	$\lambda(\text{sec}^{-1})$	$\hat{\lambda}(\text{sec}^{-1})$	n/(n+p)
100	100	0.00010	4.02×10^9	4.08×10^9	0.496
10	10	0.0109	3.9×10^4	4.6×10^4	0.462
2	1.996	0.273	9.	19.	0.330
1	0.992	1.102	0.19	0.83	0.238
0.1	0.074	182.	0	0.00109	0.130
0.08	0.058	296.	0	0.00108	0.116
0.06	0.043	535.	0	0.00107	0.089

$^+$unit = 10^{10} K.

implicitly as a function of $a(t)$ by equation (18), and of course the neutrino temperature satisfies $T_\nu \propto a(t)^{-1}$. The results of the integration are given in Table VIII-1. The neutron abundance satisfies

$$\frac{dn}{dt} = \lambda p - \hat{\lambda} n \ ,$$

where the coefficients λ and $\hat{\lambda}$ are found by numerical integration of equations like (36). Table VIII-1 differs from the similar table in ref. (7) because a somewhat different value for the coupling constant (26) was used and because the rates for $n \rightleftarrows p + e^- + \bar{\nu}$ are included here, but were counted separately in ref. (5).

Again, the main point of the Table is that there is a residual frozen-in neutron abundance ~ 0.1 to 0.2 in the temperature range from $T \cong 10^{10}$ K to $T \cong 5 \times 10^8$ K because $\hat{\lambda}t$ falls below unity in this interval. Helium is produced toward the end of the interval, as described in the next Section.

iv) *Helium Production*

The formation of heavier elements from neutrons and protons is controlled by the reaction

$$n + p \rightleftarrows d + \gamma \ . \tag{37}$$

Once an appreciable amount of deuterium accumulates it rather quickly burns to helium. We can therefore understand the orders of magnitude by studying reaction (37), and the first point clearly is the equilibrium between deuterium production and the reverse process, photo-dissociation. The computation of the equilibrium deuterium abundance is just the same as that leading to equation VII-9, the only change being that deuterium has spin 1, so we have to count three states. It is convenient to express the result in terms of abundances by number,

$$x_n = \frac{N_n}{N}, \ x_p = \frac{N_p}{N}, \ x_d = \frac{N_d}{N},$$

$$x_n + x_p + 2x_d + \dots = 1$$

where N is the total number of nucleons, free and bound, in some region of space, N_n the number of free neutrons, N_p the number of free protons, and N_d the number of free deuterons. The second equation includes the sum over particles heavier than deuterium. The equilibrium abundance ratio is

$$\left(\frac{x_n x_p}{x_d}\right)_e = \frac{4}{3}\frac{(2\pi kT)^{3/2}}{(2\pi \hbar)^3 n}\left(\frac{m_n m_p}{m_d}\right)^{3/2} e^{-B/kT} , \tag{38}$$

where $n = N/V$ is the nucleon number density and

$$B = 2.225 \text{ MeV}$$

is the binding energy of deuterium.

We can fix the nucleon density as a function of T from the present density by equations (2) and (3),

$$\begin{aligned} n &= \left(\frac{T}{2.7}\right)^3 n_0 \\ &= 5.70 \times 10^{20}\, T_9{}^3\, (\Omega h^2)\, \text{cm}^{-3}, \end{aligned} \tag{39}$$

where T_9 is the radiation temperature in units of 10^9 K, which is a convenient unit because the nuclear burning in the model happens at $T_9 \sim 1$. Equation (39) must of course be corrected for the effect of electron pair production on the temperature at $T_9 \gtrsim 10$, as in Table VIII-1.

With equation (39) equation (38) becomes

$$(x_n x_p/x_d)_e = \exp [29.23 - 25.82/T_9 - \tfrac{3}{2} \ln T_9 - \ln (\Omega h^2)] . \tag{40}$$

This equation says that as the Universe expands and cools the equilibrium shifts to favor deuterons over free neutrons and protons at $T_9 \sim 1$; and because the coefficient of $T_9{}^{-1}$ in the exponent is a large number the transition is sharp and the transition temperature not very sensitive to the density parameter Ωh^2. We can define a transition temperature T^* where the equilibrium ratio $(x_n x_p/x_d)_e$ is unity, which is where the exponent in (40) vanishes. This gives

$$T_9^* = 0.88,\ \Omega h^2 = 1,$$
$$T_9^* = 0.77,\ \Omega h^2 = 0.02\,. \tag{41}$$

According to Table VIII-1 the neutron-proton abundance ratio at temperature T^* would be $n/(n+p) \cong 0.12$ in the absence of nuclear burning, and if all these neutrons were consumed in helium production the resulting helium abundance by mass would be

$$Y = \frac{2n}{n+p} \cong 0.24\,. \tag{42}$$

As described below the detailed numerical integration of the nuclear reaction rates gives a number slightly higher than (42). This is because there is a temperature range $T_9 \cong 4.0$ to 0.8 where He^4 is thermodynamically favored over neutrons and protons but deuterium is not. Through most of this range the deuterium abundance is so very low that there is negligible helium production, but toward the end of the range the deuterium abundance, though low, is large enough to permit helium production to commence somewhat earlier than the characteristic epoch fixed by (41).

It will be noted that, by another coincidence of the numbers in the model, nuclear burning can commence just as the age of the Universe is becoming comparable to the neutron half life and the neutrons are starting to slip away through free decay.

At this point one might ask, as in Sec. VII-c-iii, whether the radiation accompanying deuterium formation perturbs the high energy tail of the radiation distribution. It is readily seen that it does not, because there is an enormous rate for destruction (and creation) of the high energy tail above 2 MeV, for example by pair creation in the field of the nuclei.

The next question is the rate of the reaction (37). At the energy of interest (~ 100 keV) the cross-section for neutron capture is inversely proportional to velocity to a good approximation, so the rate coefficient is nearly independent of temperature, and amounts to

$$\langle \sigma v \rangle = 4.55 \times 10^{-20}\ \text{cm}^3\ \text{sec}^{-1}\,. \tag{43}$$

The relevant dimensionless number is $\langle\sigma v\rangle\, n^*\, t^*$, where the star means the quantities are evaluated at the transition temperature T^* (eq. 41). If this number is large it means that almost all the neutrons that survive to 10^9 K react to form deuterium, and this in turn means that almost all the deuterium burns to helium because the cross-sections leading from deuterium to helium are relatively large. On taking the time from Table VIII-1 and the density from (39) we find

$$\langle\sigma v\rangle\, n^*\, t^* = 4000\,(\Omega h^2)\,. \tag{44}$$

If $\Omega h^2 \gtrsim 0.01$, as seems reasonable, this number is greater than 40, and the nuclear reaction rate is large enough to assure consumption of the neutrons. It is interesting, however, that by yet another numerical coincidence (44) is not enormously different from unity, and since (44) varies inversely as the cube of the present radiation temperature T_0 (eq. 39) the reaction probability would not have been large if the Universe now had been hotter than ~ 30 K.

For more precise numbers one must numerically integrate the nuclear reaction rates leading up to helium and beyond. There are two independent computations from rather different approaches, and it is a valuable check of all the details of the computation that they give closely similar answers.[7,8]

With all the assumptions in this Section there remain two free parameters. One is the density parameter, Ωh^2, which is thought to be in the range

$$1 \gtrsim \Omega h^2 \gtrsim 0.01\,. \tag{45}$$

The lower number would correspond to the usual estimate of the mass density in galaxies (eq. IV-7), $\Omega = 0.02$, with $h = 0.5$, say. Second, Tayler[13] pointed out that the weak interaction strength as estimated from the neutron half-life still is uncertain enough to contribute an appreciable uncertainty to Y.

The computed helium abundance is shown as a function of these two parameters in Figure VIII-1. The computation scheme used to find these abundances is the one described in ref. (7), but the results are almost

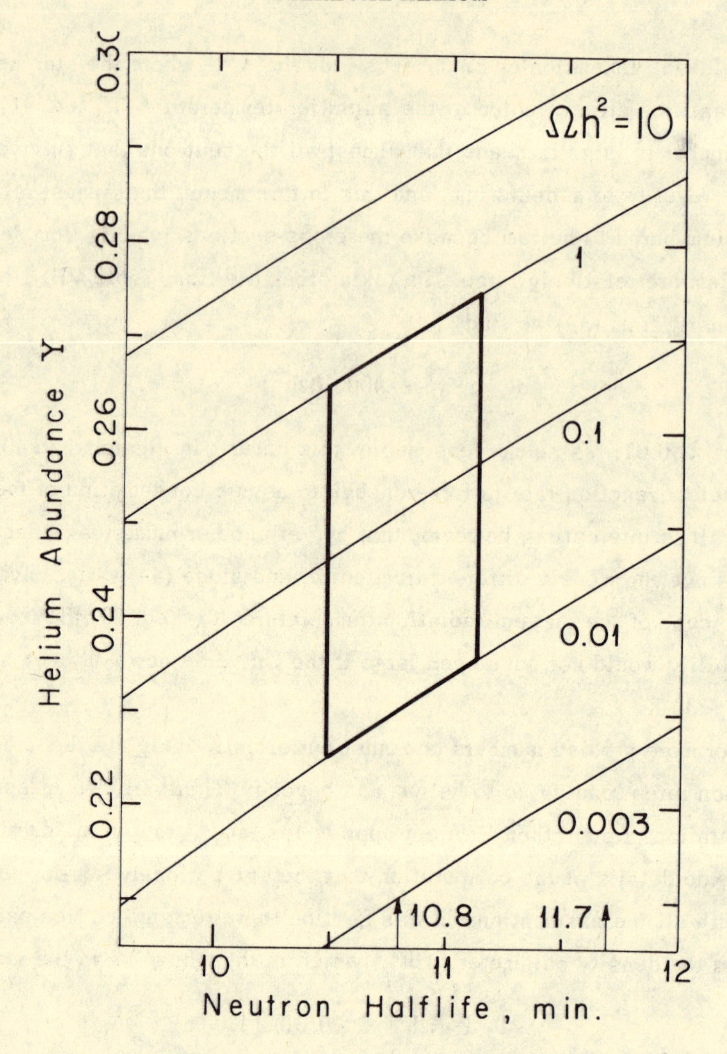

Fig. VIII-1. Computed Helium Abundance by Mass as a Function of the Density Parameter Ωh^2 and the Neutron Half-life.

indistinguishable from those of ref. (8). For example, the effective value of the neutron half-life in ref. (8) is $t_{1/2} = 12.0$ min; and the helium abundance corresponding to $T_0 = 2.7$ K, $\Omega h^2 = 1$, is given in ref. (8) as $Y = 0.289$, while Figure VIII-1 gives $Y = 0.287$. The major differences between the two computations are (1) ref. (8) took account of some 144 reactions, and one can be very sure therefore that no important path to helium (or beyond) has been missed, while in ref. (7) the attempt was to isolate the key reactions, 5 in number; (2) in ref. (8) the charged particle reaction rate coefficients $<\sigma v>$ were found by power law expansion of the cross-sections, while in ref. (7) the measured reaction cross-sections near 100 keV were fitted to a Gamow-type form $\sigma = SE^{-1} \exp(-AE^{-1/2})$, with S and A both adjustable constants, and the result numerically folded against a Maxwell-Boltzmann velocity distribution. The resulting rate coefficients differ by factors $\sim$ 2 to 4, but this affects the result hardly at all because the helium production is controlled mainly by the simple reaction (37), which anyway to first approximation only acts like a switch because (41) is appreciably larger than unity.

The residual deuterium and He^3 abundances are more sensitive to these details, but the two computations again give reasonably consistent results. The deuterium and He^3 abundances by mass are shown as

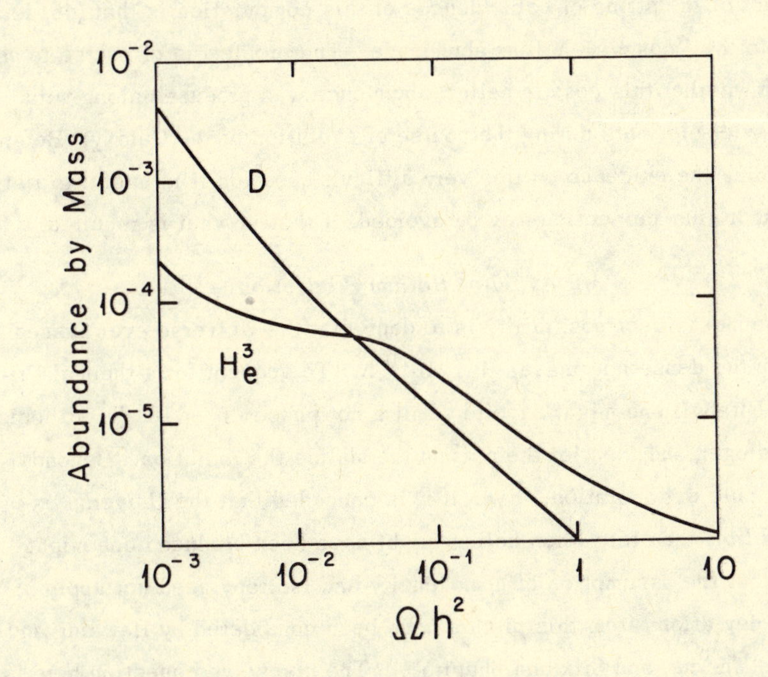

Fig. VIII-2. Computed Abundances by Mass of Deuterium and Helium^3.

functions of the parameter Ωh^2 in Figure VIII-2, with $t_{1/2} = 10.8$ min. By comparison the He^3 abundance by mass in the original material of the Solar System as deduced from the "primeval gas" found in meteorites is $\sim 6 \times 10^{-5}$, and the deuterium abundance by mass $\sim 10^{-4}$. These cannot be directly compared to the computed results in Figure VIII-2, however, because He^3 and deuterium can be created and destroyed, burned to He^4, in the cycling of material through stars.

The box in Figure VIII-1 is centered on the neutron half-life (27), and the width of the box is two standard deviations either way from the central value. The older value for the neutron half-life, 11.7 min., is indicated toward the right hand side of the figure. The top and bottom of the box correspond to the mass parameter limits (45). The range of helium abundance by mass admitted by the box is

$$0.225 \leq Y \leq 0.275 . \tag{46}$$

The last of the string of coincidences of this computation is that (46) is close to the "cosmic" helium abundance. The problem is of course to establish whether this cosmic helium abundance was produced along with the heavier elements during the course of evolution of the Galaxy. Before describing the evidence on this very difficult question it is useful to list how the helium production may be avoided, if that is what is required.

b) *Avoiding Helium Production*

The most direct possibility is to deny that the Universe ever passed through the dense hot phase, $T > 10^{10}$ K. To account for a thermal Primeval Fireball one might have to admit a hot phase $T \sim 10^4$ K, to ionize the hydrogen and then let the plasma thermalize the radiation. Beyond that is pure extrapolation. Even if it is conceded that the Universe expanded from the state when helium might have been produced one might argue that the assumption of homogeneity and isotropy need not apply. Strong deviations from this picture have been considered by Hawking and Tayler, Thorne, and Silk and Shapiro.[14] The unresolved question here is whether the Universe is truly stable against such perturbations. A more modest departure suggested by Harrison[15] that would not raise questions of stability prior to recombination would be to accept that the radiation (and space curvature) are smooth and isotropic, but that the matter is placed in some irregular fashion. Because the matter contributes negligible mass density compared to the radiation the expansion rate would be unaffected, and the radiation drag would prevent the matter from collecting in bound lumps until $z \sim 1000$. Where the matter density is low the number (44) would be reduced, so one could reduce the helium production in some fraction of the matter. However the amount of pure hydrogen one could save in this way is small. For example, let us consider the most favorable case for this idea, $\Omega h^2 = 0.01$. If a fraction $1 - \epsilon$ of the matter were concentrated in lumps, the remaining fraction ϵ smoothly distributed between the lumps, the characteristic number (44) in the inter-lump material would be 40ϵ. If we chose $\epsilon = 0.01$ the helium production in the inter-lump material would be reduced to ~ 2 percent by mass, but there would also be ~ 2 percent by mass deuterium, which seems much too high to be lowered to the present value by cycling through stars. That is, one can save some nearly pure hydrogen, but much less than one percent of the mass in galaxies, which hardly seems worthwhile.

Even if it were granted that the Universe has expanded in an isotropic and homogeneous way from $T \sim 10^{11}$ K one might question the application of ordinary general relativity to get the expansion rate. An alternative is the scalar-tensor theory of Brans and Dicke,[16] in which the effective value of Newton's constant G would have been larger in the past, and the scalar field would contribute to the total mass density, both effects helping to speed the rate of expansion (which goes as $(G\rho)^{1/2}$). This effect has been discussed by Dicke[9] and by Greenstein.[17] Dicke concludes that the speed-up factor for expansion through $T \sim 10^{10} - 10^{11}$ K is not unambiguously fixed in the theory because there are more available parameters (initial values) than one can fix from knowledge of present conditions. He gets speed-up factors ranging from ~ 2.2 up to very large values. The result of a modest speed-up like a factor of ~ 2 is to increase the helium abundance to ~ 40 percent. The helium abundance increases because (44) still is large, so the neutrons still all are consumed, but the neutron abundance is frozen in earlier, at a higher temperature, corresponding to a larger value of n/p. It is not enough to speed up the expansion to the point that (44) is on the order of unity, adjusted so that ~ 25 percent of the matter is consumed in the reaction (37), because an appreciable fraction of the deuterium thus formed gets hung up as deuterium and He^3, making unreasonably large abundances of these isotopes. If $\Omega h^2 \sim 1$ the expansion rate must be speeded up by a factor $\gtrsim 10^6$ over the Lemaître model to make low helium and acceptably low deuterium and He^3 abundances.[7] On the other hand it is amusing to note that if the expansion were *slowed* by a factor of 10 it would reduce the primeval helium production to $Y < 0.10$, because the neutron abundance is frozen in at a lower temperature and the neutrons have longer to decay before nuclear burning commences.

If the lepton number of the Universe is large enough in magnitude the chemical potential (Fermi energy) of the neutrinos cannot be neglected, and the equilibrium n/p abundance ratio (19) must be modified.[18] This was discussed in detail by Wagoner, Fowler and Hoyle.[8] To fix orders of magnitude we will consider here only the two limiting cases, where the Fermi energy of neutrinos or of antineutrinos $>> kT$, so that the neutrinos or antineutrinos are degenerate.

When the Fermi energy is E_f the neutrino (or antineutrino) number density is (eq. 9)

$$n_\nu = \frac{4\pi}{3}(E_f/hc)^3 \ , \qquad (47)$$

and the degeneracy condition is

$$E_f >> kT \ . \qquad (48)$$

It will be noted that if the neutrinos are degenerate the expansion of the Universe preserves the degeneracy, for we know $n_\nu \propto a(t)^{-3}$, so (47) says $E_f \propto a(t)^{-1}$, but we also know that $T \propto a(t)^{-1}$. With $T_0 = 2.7$ K the present Fermi energy (48) for degeneracy would be

$$E_f >> kT_0 \sim 2 \times 10^{-4} \text{ eV} ,$$

distressingly hard to detect (ref. IV-72). The present degenerate neutrino (or antineutrino) number density would be

$$n_\nu(t_0) >> 30 \text{ cm}^{-3} . \tag{49}$$

If there were a degenerate neutrino sea the reaction

$$e^- + p \rightarrow n + \nu$$

would be cut off because there is no room for the neutrino. One might think to make neutrons by

$$\bar{\nu} + p \rightarrow n + e^+,$$

but of course as neutrino pair production is cut off there are no antineutrinos. On the other hand there are ample neutrinos to eliminate neutrons by $\nu + n \rightarrow p + e^-$. There being no neutrons the primeval helium production is neatly eliminated.

If the antineutrinos are degenerate and if the degeneracy energy $E_f > Q - m_e c^2$ (eq. 19) one finds as above that there are only neutrons. As soon as E_f falls below $Q - m_e c^2$ the neutrons can start to decay. If this happens too soon neutrons and protons can react, and we end up with too much helium or deuterium. To fix the orders of magnitude we can suppose that the conversion from neutrons to protons takes place in one free neutron half-life, $t_{1/2} \sim 10$ min. Then we require that, at the epoch where E_f falls to 0.78 MeV,

$$\langle \sigma v \rangle \, n \, t_{1/2} << 1 .$$

With equations (39) and (43) this says that the neutrons should not decay before the epoch

$$T_d << 4 \times 10^7 \, (\Omega h^2)^{-1/3} \text{ K} .$$

As the degeneracy energy at T_d is 0.78 MeV the present antineutrino degeneracy energy would be

$$E_f = (Q - m_e c^2)\, T_0/T_d >> 0.05\, (\Omega h^2)^{1/3}\ \text{eV}\ ,$$

and the present equivalent mass density in antineutrinos would be

$$\rho_\nu >> 4 \times 10^{-26}\, (\Omega h^2)^{4/3}\ \text{g cm}^{-3}\ .$$

By comparison conventional cosmological models would only permit

$$\rho_\nu \lesssim 10^{-28}\, h^2\ \text{g cm}^{-3}\ .$$

To summarize, a degenerate antineutrino sea with mass density consistent with the conventional cosmological models would be catastrophic because it would cause all the hydrogen to be burned. A degenerate neutrino sea would preserve pure hydrogen. However, one must be prepared to admit that the lepton number of the Universe is more than seven orders of magnitude larger than the nucleon number. As we have no direct observational measure of the ratio it is a matter of philosophy whether one considers that this is a distressing admission to have to make.[18,19]

Before the discovery of the microwave background Zeldovich[18] pointed out that helium production in the Big Bang would be avoided if the Universe started out cold, at zero entropy, with equal number densities n of protons, electrons and neutrinos. Then once the degeneracy energy of the nucleons falls well below $m_n c^2$ the equilibrium favors elimination of the neutrons, for to make a neutron by electron capture requires energy expenditure $Q + hc\left(\frac{3n}{4\pi}\right)^{1/3}$, where the second term is the neutrino degeneracy energy, but the available energy from the top of the electron sea is $(m_e^2 c^4 + h^2 c^2 (3n/8\pi)^{2/3})^{1/2}$, which always falls short because of the extra factor of 2 from the two electron spin states. Thus the neutrons are eliminated. Once the neutrino degeneracy energy falls below $Q_d = 0.42$ MeV, deuterium can form by the usual reaction in stars,

$$p + p \rightarrow d + e^+ + \nu\ .$$

The proton density when this reaction can commence is

$$n = \frac{8\pi}{3} (Q_d/hc)^3$$

$$= 3 \times 10^{29} \text{ cm}^{-3} ,$$

and the age of the cosmological model is

$$t = (6\pi \, Gm_n n)^{-1/2}$$

$$= 1 \text{ sec.}$$

The cross-section for this process being $\sim 10^{-49}$ cm^2, we can conclude that the matter would remain pure hydrogen.

There is yet another interesting variant of the neutrino problem. We have assumed that there are two species of neutrinos belonging to the electron-like particles, e and μ. If there were other heavier electrons with associated massless neutrinos, these new neutrinos would increase the expansion rate, hence increase the residual neutron abundance. Each class of massless neutrinos, with one spin state, would contribute to the present mass density

$$\rho_\nu(t_0) = \frac{7}{8}\left(\frac{4}{11}\right)^{4/3} aT_0{}^4/c^2 .$$

The total neutrino density could be as large as

$$\rho_T \lesssim 10^{-29} \text{ g cm}^{-3}$$

without seriously affecting the cosmological models. This could increase the expansion rate through the helium production epoch by the factor

$$\left[\frac{\rho_T c^2}{aT_0{}^4}\right]^{1/2} \lesssim 150 .$$

Even in the extreme low density model, $\Omega h^2 \sim 0.01$, this speed-up factor would not be enough to eliminate excessive deuterium production (eq. 44; ref. 7). That is, extra classes of neutrinos could only cause trouble for the model.

Our list of possibilities is far from complete. We have not discussed the thought that constants other than the strength of the gravitational interaction may vary with time, for example, or the thought that matter may be issuing from nuclei of galaxies.

c) *Comparison with the Observed Helium Abundances*

The problem here is to compare the computed primeval helium abundance (46) with the observed cosmic abundance. If the helium abundance in some object is found to agree with (46) it does not confirm the theoretical prediction, for the helium might have been produced along with the heavier elements in earlier generations of stars. If this were so one would look for a variation in helium abundance depending on age and position (whether the object is in the galactic nucleus, or the disc, or the halo, or wherever) as is observed for the heavy element abundance. If the helium abundance is found to be constant under a wide range of objects it will be strong evidence for some universal synthesis process like the neutron burning in the Big Bang. On the other hand, if we can find an object that has helium abundance clearly less than (46), and if we can convince ourselves that the helium content has not been depleted somehow, the "standard" cosmology on which (46) is based is in trouble.

A careful survey of the cosmic helium problem has been given by Danziger.[20] The following are the main lines of attack, more or less in order of increasing age of the objects studied.

1) *Optical Recombination Lines in* H II *Regions.* An H II region is the ionized interstellar gas around stars hot enough to be emitting ionizing ultraviolet radiation. One looks for the recombination lines of hydrogen and helium, and evidently the ratio of line strengths varies as the abundance ratio $N(H_e^+)/N(H^+)$ and as the ratio of recombination rate coefficients $<\sigma v>$. As it happens this latter ratio is not very sensitive to the plasma temperature. The abundance ratio has to be corrected for an estimate of the abundance of neutral helium, the correction typically being $\lesssim$ 10 percent.

TABLE VIII-2

OPTICAL HELIUM ABUNDANCES IN H II REGIONS

	Y
Mean for the Galaxy	0.29
S.M.C.	0.25
L.M.C.	0.29
M33	0.34
NGC 6822	0.27
NGC 4449	0.28
NGC 5461	0.28
NGC 5471	0.28
NGC 7679	0.29

Some helium abundances found in this way are listed in Table VIII-2. These numbers are taken from Peimbert and Spinrad, and include data from a number of workers.[21] The helium abundance by mass, Y, is related to the abundance ratio of helium to hydrogen by number, n_{He}/n_H, by the formula

$$Y = \frac{(1 - Z)\, 4n_{He}/n_H}{1 + 4n_{He}/n_H} ,$$

where Z is the abundance by mass of elements heavier than helium. Here and throughout we ignore the small correction for Z ($\lesssim$ 5 percent) and convert according to the formula

$$Y = \frac{4n_{He}/n_H}{1 + 4n_{He}/n_H} .$$

The first number in the table is the mean for several H II regions in the Galaxy. The next four numbers are for H II regions in other galaxies in the Local Group, and the last four numbers for galaxies outside the Local Group. Within the uncertainties of the measurements there is no evidence of real scatter in the helium abundances. However, Ford and Rubin have found evidence that there is a real and appreciable variation of helium strength in the emission nebulae in M 31.[22]

2) *Radio Recombination Lines.* In the very dilute plasma in an interstellar emission nebula an electron can recombine to a level of large principal quantum number $n \sim 100$ and $\ell \sim n$ in hydrogen, and the radiation emitted in the transition $n \rightarrow n-1$ then has radio frequency. This very useful and important effect was first discussed in detail by Kardashev.[23] He remarked also that one can distinguish the radio frequency recombination lines of hydrogen and helium (He^+ with an electron in a large n, hydrogen-like orbit) because the center of mass corrections are different. Assuming the hydrogen and helium are well mixed in the plasma and the optical depth is small the ratio of observed integrated line strengths is equal to the abundance ratio $N(He^+)/N(H^+)$. The helium line was first detected by Lilley, Palmer, Penfield and Zuckerman, in 1966.[23] The recent results are in reasonable agreement with the optical data, and a mean value for emission regions in the Galaxy is[23,24]

$$Y(\text{H II-Radio}) \sim 0.26 \,. \tag{50}$$

The abundance seems to be nearly constant from source to source in the Galaxy with the exception of the double source in Sagittarius, near the galactic center, where the helium line is not observed and it should be if the abundance were as high as (50). There is always the possibility that the helium is not ionized because the exciting stars are not hot enough, but this is judged to be unlikely.[24] Any further judgment must await better understanding of this source.

3) *Pop* I *Stellar Spectroscopic Abundances.* Young massive 0 and B stars are hot enough to show helium lines, and these may be used with a model atmosphere to get a helium abundance. In a recent survey Shipman and Strom[25] concluded that the spectroscopic helium abundance in young stars is roughly consistent with the constant value

$$Y(\text{Pop I spectroscopic}) \sim 0.28 \,. \tag{51}$$

4) *Pop* I *Main Sequence.* The relation between stellar mass and luminosity on the main sequence, which is to say before the star has appreciably burned out its central parts, depends on the helium and heavy element abundances, Y and Z. For given stellar mass and given Z the luminosity increases with increasing Y because larger Y means larger mean molecular weight, hence higher temperature needed to hold up the star, hence higher temperature gradient, hence greater heat flux out of the star. This test was discussed by Percy and Demarque and by Morton; and a recent study by Popper, Jørgensen, Morton and Leckrone gave[26]

$$Y\ (\text{POP I M.S.}) \sim 0.32\ . \tag{52}$$

5) *Solar Abundance – Spectroscopic and Solar Cosmic Rays.* The Sun is about 4.6×10^9 y old, much older than the objects discussed above and perhaps half as old as the oldest stars in the Galaxy. One cannot get the solar helium abundance from the observed emission lines because these originate in the corona where the situation seems to be too complicated for any convincing interpretation. One can arrive at Y by two steps: One can get the abundance ratio He/CNO from the abundance in solar cosmic rays. Because fully ionized helium and medium weight nuclei have very nearly the same charge-to-mass ratio they should be treated the same way in the electromagnetic processes that accelerate the particles, and indeed the abundance ratio does seem to be reasonably constant from event to event, in distinction to the He/H ratio. Second, one can get the ratio CNO/H from spectroscopic observations. The product of these two numbers is He/H. Gaustad first applied this method,[27] and a more recent result is[28]

$$Y\ (\text{Solar C.R.}) \cong 0.20\ . \tag{53}$$

It will be noted that this falls just below the range of equation (46).

6) *Solar Abundance – Solar Models.* There has been extensive rediscussion of the Solar models because of their failure to account for the null result of the Davis solar neutrino experiment. A recent model by Bahcall and Ulrich gave[29]

$$Y\ (\text{Solar model}) \cong 0.26\ . \qquad (54)$$

This abundance is larger than (53), and consistent with (46). The Bahcall-Ulrich model, like other solar models, predicts a solar neutrino flux greater than Davis's upper limit.

7) *Pop* II *Spectra*. We come now to the old Population II, whose members do show pronounced deficiencies of heavy element abundances, suggesting that they are closer to the primordial composition. Unhappily the Pop II stars we see are still shining because they have low mass and so evolve slowly, and generally with the low mass goes a surface temperature too low to excite helium lines. However, when an old star burns out the hydrogen in its central parts it can for a short time become a hot bright horizontal branch blue star in which helium lines are excited. In some cases the helium lines in these stars are much weaker than in Pop I stars with the same surface temperature, which might suggest that the older stars are deficient in helium by a factor ~ 10 to 100.[30] However, Sargent and Searle found that at least in some cases where the helium lines in the Pop II stars were "anomalously" weak there were other spectral peculiarities such as strong phosphorus lines which also are found in some Pop I stars which have unusually weak helium lines. Apparently until these anomalies are understood it will be hard to relate the weak helium lines to the helium abundance of the whole star.[31]

8) *Pop* II *Stellar Models*. Models for the structure and evolution of the Pop II stars have to account for a wealth of observational data, and the hope is that the fit of the model to all these details will fix as a parameter the initial helium abundance. Christy's models for R. R. Lyrae variable stars seem to be best fitted with helium abundance[32]

$$Y\ (\text{R. R. Lyrae}) \cong 0.32\ . \qquad (55)$$

Faulkner, Iben, Demarque and others recently have argued that the older stars must have high helium abundance, $Y \sim 0.3$, to account for their distribution in the Hertzsprung-Russel diagram, but in view of the great uncertainties like mass loss and mixing it is hard to know how sure this is.

9) *Pop* II *Mass-Luminosity Relation.* One would like to play game (4) for the Pop II, but unhappily there are no well-determined masses for Pop II stars. The star μ Cassiopeia is an interesting candidate because it is mildly deficient in heavy elements relative to the Sun, we know it has a dark companion because it is observed to wiggle in the sky, and if the separation of the two stars, which is thought to be $\sim 1''$ arc, could be measured one could get the mass from Kepler's laws.[33] Hegyi and Curott attempted to measure the separation, and arrived at tentative evidence possibly favoring low helium abundance.[34]

10) *Pop* II *Planetary Nebula.* As was mentioned above there is one planetary nebula in a Pop II system, the globular star cluster M15. O'Dell, Peimbert and Kinman[5] found that the oxygen abundance in the nebula is markedly deficient relative to Pop I, as one might expect, but that the helium abundance is

$$Y \text{ (M15 Planetary)} \sim 0.4 .$$

Since this is material ejected from a highly evolved and active star one cannot be entirely sure how much of this helium was formed in the star.

11) *Quasars.* Assuming the cosmological interpretation of the redshifts of the quasi-stellar objects, or quasars, the observed quasars span a range of cosmic time from fairly recent to something comparable to the age of the galactic Pop II. It is quite unclear whether quasars are a pathology in the course of evolution of galaxies, or are normally associated with the formation of galaxies, or are something quite different. Apparently, it was first pointed out by Osterbrock and Parker that the helium emission lines in some quasars are weak, and this has been verified by a number of people.[21,35] The detailed model of Bahcall and Kozlovsky for the quasar 3C 273 gave a helium abundance a factor of 10 below the Solar value, which is to say well below the range of values admitted in Figure VIII-1. On the other hand there were roughly "normal" abundances of heavier elements. On the face of it this is strong evidence *against* the primeval helium production picture. However, until we have a deeper understanding of the nature of quasars it will be hard to argue that the evidence is conclusive.

d) *Discussion*

We would like to understand what the observations are trying to tell us about the helium abundance, and then make a clean statement of what is the theoretical significance of this revelation. On both fronts the situation is at best cloudy. On the observational side a central point is that the helium abundance exhibits uniformity in the face of diversity, as a number of observers have been remarking for a number of years. The issue is whether there are exceptions to the rule, objects that have low helium abundance because they are made of material that is relatively untouched by the fruits of stellar evolution, and that never contained much helium. At the moment the best candidates are the quasi-stellar objects and the H II source near the center of the Galaxy, but in neither case is the object very well understood. As is so often the case in astrophysics no one observational result can settle the matter, and it will be a question of lengthy debate on the accumulated weight of the evidence. In the primeval helium problem it is too soon to measure the balance.

On the theoretical side we are in any case faced with the task of understanding the fairly uniform cosmic helium abundance. The solution tentatively favored by some is that the helium originated in supermassive objects, a sort of Pop III.[6,8,36] The problem here is to understand how the process could have been controlled so well – how the objects were prevented from eliminating most or all of the hydrogen, for example, or how the large scale composition was kept so uniform.

What of helium production in the Big Bang? We have shown that there are a number of very explicit ways in which the calculation of the final result (46) can fail, and any number of vague fears. Therefore one should look at this problem from quite the opposite point of view. It is absurd on the face of it to think that one could ever hope to predict the primeval helium abundance. Our understanding of the Universe as it is now is so weak that we would hesitate to say what the Universe was like at a redshift of 3, let us say, yet we propose to analyze the Universe at a redshift of 3×10^9! To my mind there are four points that may justify including this enormous theoretical flight in a review of observational cosmology. The first point is that it is the straightforward extrapolation from a commonly accepted and modestly successful picture, and one ought to follow up straight consequences. The second is the possible discovery of the Primeval Fireball, which encourages us to believe in a Big Bang of some sort. The third is the fact that the helium abundance commonly is large, around 30 percent by mass, by a miraculous coincidence comparable to the prediction of the naive Big Bang. The final point is the purely theoretical view that the Universe seems to be unstable. If this is so it places an impressively strong constraint on what the early Universe must have been like, so that we can use symmetry arguments to compute under a finite number of assumptions.

It is not very profitable to list ways by which these points may be negated, or the primeval helium production otherwise eliminated, because it is so easy to eliminate the helium. That is, if we can show that the Galaxy started out with little or no helium we will not have learned much of theoretical interest to cosmology, only that a very specific chain of events did not happen. An exciting prospect is that we may eventually find that the Galaxy started out with high helium, in precise agreement with the number from the naive Big Bang, which would be a positive result we could with justification tie to the definite model. Of course, we may be confronted with a primeval helium abundance $Y = 0.1$, let us say.

REFERENCES

1. This calculation is described by R. A. Alpher and R. C. Herman, *Rev. Modern Phys.* 22, 153, 1950.
2. C. Hayashi, *Progr. Theor. Phys.* 5, 224, 1950.
3. R. A. Alpher, J. W. Follin and R. C. Herman, *Phys. Rev.* 92, 1347, 1953.
4. D. E. Osterbrock and J. B. Rogerson, *P.A.S.P.* 73, 129, 1961.
5. C. R. O'Dell, M. Peimbert and T. D. Kinman, *Ap. J.* 140, 119, 1964.

6. F. Hoyle and R. J. Tayler, *Nature* 203, 1108, 1964.
7. P. J. E. Peebles, *Ap. J.* 146, 542, 1966.
8. R. V. Wagoner, W. A. Fowler and F. Hoyle, *Ap. J.* 148, 3, 1967.
9. R. H. Dicke, *Ap. J.* 152, 1, 1968.
10. R. F. O'Connell and J. J. Matese, *Nature* 222, 649, 1969; *Ap. J.* 160, 451, 1970; G. Greenstein, *Nature* 223, 938, 1969.
11. L. Landau and E. Lifshitz, *Statistical Physics* (Addison-Wesley), 1958, §104. A convenient list of the integrals in equations like (11) and (13) is given in §57.
12. C. J. Christensen, A. Nielson, A. Bahnsen, W. K. Brown and B. H. Rustad, *Phys. Letters* 26B, 11, 1967.
13. R. J. Tayler, *Nature* 217, 433, 1968.
14. S. W. Hawking and R. J. Tayler, *Nature* 209, 1278, 1966; K. S. Thorne, *Ap. J.* 148, 51, 1967; J. Silk and S. Shapiro, not published.
15. E. R. Harrison, *A. J.* 73, 535, 1968.
16. C. Brans and R. H. Dicke, *Phys. Rev.* 124, 925, 1961.
17. G. Greenstein, *Astrophysics and Space Physics* 2, 155, 1968.
18. Ya B. Zeldovich, *Zh. E.T.F.* 43, 1561, 1962; Engl. tr. *Soviet Physics - J.E.T.P.* 16, 1102, 1963; *Usp. Fig. Nauk* 80, 357, 1963; Engl. tr. *Soviet Physics - Uspekhi* 6, 475, 1964; R. H. Dicke, P. J. E. Peebles, P. G. Roll and D. T. Wilkinson, *Ap. J.* 142, 414, 1965.
19. W. A. Fowler, *Comments on Astrophysics and Space Physics* 2, 134, 1970.
20. I. J. Danziger, *Ann. Rev. Astron. Astrophys.* 8, 161, 1970.
21. M. Peimbert and H. Spinrad, *Ap. J.* 159, 809, 1970; *Astron. and Astrophys.* 7, 311, 1970.
22. W. K. Ford and V. C. Rubin, *B.A.A.S.* 1, 188, 1969.
23. N. S. Kardashev, *Astron. Zh.* 36, 838, 1959; Engl. Trans. *Soviet Astronomy - A.J.* 3, 813, 1960; P. Palmer, B. Zuckerman, H. Penfield, A. E. Lilley and P. G. Mezger, *Nature* 211, 174, 1966; *Ap. J.* 156, 887, 1969; M. A. Gordon and E. Churchwell, *Astron. and Astrophys.* 9, 307, 1970.

24. P. G. Mezger, T. L. Wilson, F. F. Gardner and D. K. Milne, *Astrophysical Letters* 6, 35, 1970.

25. H. L. Shipman and S. E. Strom, *Ap. J.* 159, 183, 1970.

26. J. R. Percy and P. Demarque, *Ap. J.* 147, 1200, 1967; D. C. Morton, *Ap. J.* 151, 285, 1968; D. M. Popper, H. E. Jørgensen, D. C. Morton and D. S. Leckrone, *Ap. J.* 161, L57, 1970.

27. J. Gaustad, *Ap. J.* 139, 406, 1964.

28. N. Durgaprasad, C. E. Fichtel, D. E. Guss and D. V. Reames, *Ap. J.* 154, 307, 1968.

29. J. N. Bahcall and R. K. Ulrich, *Ap. J.* 160, L57, 1970.

30. J. L. Greenstein and G. Münch, *Ap. J.* 146, 618, 1966; W. L. W. Sargent and L. Searle, *Ap. J.* 145, 652, 1966.

31. W. L. W. Sargent and L. Searle, *Ap. J.* 150, L33, 1967.

32. R. F. Christy, *Ap. J.* 144, 108, 1966; A. Sandage, *Ap. J.* 157, 515, 1969.

33. T. R. Dennis, *P.A.S.P.* 77, 283, 1965.

34. D. Hegyi and D. Curott, *Phys. Rev. Letters* 24, 415, 1970.

35. D. E. Osterbrock and R. A. Parker, *Ap. J.* 143, 268, 1966; J. N. Bahcall and B. Kozlovsky, *Ap. J.* 155, 1077, 1969.

36. G. Burbidge, *Comments in Astrophysics and Space Physics*, 1, 101, 1969.

APPENDIX

NOTATION, CONVENTIONS AND UNITS

Notation

The following symbols are used often enough to bear listing.

$a = a(t)$ is the expansion parameter, the scale factor which measures the expansion of the Universe. On occasion a also is Stefan's constant.

$a_0 = a(t_0)$ is the present value of the expansion parameter.

H is the present measured value of Hubble's constant. The expansion rate at any earlier epoch is written $\dot{a}/a$.

h is Hubble's constant in units of $100 \text{ km sec}^{-1} \text{ Mpc}^{-1}$; and occasionally h is Planck's constant.

q is the acceleration parameter, measured now; not a function of time.

z is the cosmological redshift, where $1 + z$ is the ratio of observed wavelength to wavelength measured at the source.

t is real physical time, as kept by a comoving observer, and generally understood to be measured from $t = 0$ at the singular point of infinite density in the Big Bang models.

t_0 is the present value of t.

ρ is the mean mass density, generally a function of time.

$\rho_0 = \rho(G) + \rho(H\ I) + \ldots$ is the present mean mass density, a sum of the contributions from galaxies, uniformly distributed atomic hydrogen, etc.

$\rho_c = 3\ H^2/(8\pi\ G)$ is the present mass density in the Einstein-deSitter model.

i is radiation brightness, energy flux per steradian. A subscript or argument ν means brightness per frequency interval, subscript or argument λ brightness per wavelength interval.

f represents an energy or particle flux; $\mathcal{L}$ is the luminosity of an object, rate of radiation of energy.

T is the Primeval Fireball temperature, a function of time.

$T_0 \cong 2.7$ K is the present Fireball temperature.

T_e is the electron or matter temperature.

T_a is an antenna temperature, a linear measure of i_ν.

ℓ is a real physical distance, measured say in cm or Mpc. In principle ℓ is the result of a measurement with a very long piece of string.

r is coordinate position in the Lemaître line element (eq. I-2).

The Greek indices $\alpha, \beta, \ldots$ on the components of a tensor range from 1 to 3, Latin indices $i, j \ldots$ from 0 to 3. We use the standard summation convention that repeated indices are supposed to be summed over the range of the index. A partial derivative occasionally is represented by a comma,

$$\frac{\partial \phi}{\partial X^\alpha} = \phi_{,\alpha} ,$$

and a derivative with respect to time by a dot,

$$\dot{a} = \frac{da}{dt} .$$

Natural logarithms are written

$$\ln (X) ,$$

common logarithms, to base 10, are written

$$\log (X) .$$

Parsecs and Magnitudes

A very useful guide to measures, orders of magnitude, and typical values in astronomy is C. W. Allen's Astrophysical Quantities.[1] There are two measures used in these notes so frequently as to justify an explicit definition here.

We have adopted as the standard astronomical unit of length the parsec, which is the distance at which one second of arc subtends one Astronomical Unit = mean Earth-Sun distance. The conversion factors are

$$1 \text{ pc} = 3.086 \times 10^{18} \text{ cm}$$
$$= 3.26 \text{ light years.} \tag{1}$$

In extragalactic astronomy the more convenient number is one million parsecs,

$$1 \text{ Mpc} = 3.086 \times 10^{24} \text{ cm} . \tag{2}$$

The observed brightness in the sky of a star or galaxy is expressed in a logarithmic scale. Let f_1 be the incident energy flux from an object, ergs cm^{-2} sec^{-1} in some chosen wavelength band, and let f_2 be the energy flux from a second object in the same band. Then the apparent magnitudes of the two objects are related by the equation

$$m_2 - m_1 = 2.5 \log (f_1/f_2) . \tag{3}$$

It will be noted that the dimmer the object the greater the apparent magnitude.

The incident flux f should be corrected for the effect of space absorption and scattering within the Galaxy, but this is a refinement we need not bother with. m will be understood to be the corrected apparent magnitude.

Until fairly recently the standard measure of brightness of a galaxy was the photographic magnitude system, which refers to the incident energy flux in a band centered roughly on 4300 A wavelength. Apparent magnitudes measured in this system carry the subscript pg. A more modern system centered on about the same color is the blue magnitude, m_B. The common practice now is to use a magnitude system centered on about 5500 A wavelength, denoted m_V. The magnitude difference $m_B - m_V = B - V$ is a measure of the color of the object, independent of distance (if absorption and redshift may be neglected). The zero points of the magnitude systems are such that the color index $B - V$ typically is ~ 0.9 for galaxies.

The zero points of the magnitude systems are fixed by the equations, from Allen, and Matthews and Sandage,[2]

$$\log(f_\lambda) = -0.4\, m_V - 8.42, \quad \lambda = 5500 \text{ A},$$
$$\log(f_\lambda) = -0.4\, m_{pg} - 8.3, \quad \lambda = 4300 \text{ A}, \tag{4}$$

where f_λ has units of ergs cm^{-2} sec^{-1} A^{-1}. Because the magnitudes refer to energy flux in rather broad bands the conversion factor depends somewhat on the spectrum.

In some circles it is the practice to express surface brightness in units of $S_{10}(V)$, which is the equivalent number of 10th magnitude visual stars per square degree. By equation (4) the conversion factor is

$$\nu i_\nu = \lambda i_\lambda = 6.86 \times 10^{-6}\, S_{10}(V) \text{ ergs cm}^{-2} \text{ sec}^{-1} \text{ ster}^{-1},$$
$$\lambda = c/\nu = 5500 \text{ A}. \tag{5}$$

The absolute magnitude M of an object is a measure of its luminosity, ergs sec^{-1} radiated energy in a chosen wavelength band, on the same logarithmic scale as equation (3). The normalization of M is such that if the object were placed at the fiducial distance 10 pc from us its apparent magnitude would be equal to its absolute magnitude M. If the true distance is ℓ then by the inverse square law the incident flux is smaller than it would be for the fiducial distance, by the factor $(\ell/10 \text{ pc})^2$, so by equation (3) its apparent magnitude at distance ℓ satisfies

$$m - M = 5 \log(\ell/10 \text{ pc}). \tag{6}$$

$m - M$ is called the distance modulus.

A convenient reference point is that the absolute magnitude of the Sun is $M_V = 4.79$. Therefore the luminosity of an object with absolute magnitude M_V is

$$\mathcal{L}(M_V) = 10^{0.4\,(4.79 - M_V)}\, \mathcal{L}_\odot, \tag{7}$$

the comparison with the solar luminosity being at about 5500 A wavelength. The brightest normal galaxies are at $M_V \sim -22$, which translates to $\sim 5 \times 10^{10} \mathcal{L}_{\odot}$.

REFERENCES

1. C. W. Allen, *Astrophysical Quantities,* (Athlone Press, London) 2nd ed., 1963.
2. T. A. Matthews and A. R. Sandage, *Ap. J.* 138, 30, 1963.